STEAM ENGINES AND WATERWHEELS

a pictorial study of some early mining machines

STEAM ENGINES

and

WATERWHEELS

a pictorial study of some early mining machines

Frank D. Woodall

MOORLAND PUBLISHING COMPANY

ISBN 0 903485 35 4

Printed in Great Britain by
Wood Mitchell & Co Ltd, Stoke on Trent

For the Publishers
Moorland Publishing Company
The Market Place, Hartington,
Buxton, Derbys, SK17 0AL

Contents

Illustrations

1 Introduction

In recent years many books on mines in Cornwall, the northern Pennines, Wales, Furness and other places have been published. Nearly all of them have involved scholarly and painstaking research, but have dealt with mines from a social and economic viewpoint. In contrast this book does not deal with the miners or mine owners, nor wages and profits, but it considers the equipment used in the mines. Emphasis is placed on the larger items of machinery, such as pumping and winding engines, together with equipment for primary crushing. To leave out crushing would eliminate several interesting items, while to go further would encroach on other lines of study.

The photographs presented here show much detail because they were originally taken to help in building engineering models. Nearly all the illustrations are from the writer's own negatives, a few others being copies of old photographs from various sources. Photography has been practised for so long that there ought to be many photographs of old engines taken when they were new. Where are these old pictures? Many must be locked away in museum store rooms where there is no staff to sort them. Far more will have been destroyed by house-proud widows when having a thorough clear out and making everything clean and tidy. It is hoped that this collection of previously unpublished photographs will help to unearth other old and historically valuable illustrations.

The great copper and tin mines of Cornwall have long been noted for their fine beam engines, but it was in the neighbouring china clay works that examples of Victorian engineering at its finest remained in use for so long. Much of this equipment was secondhand, having come from the nearby metal mines. Since the china clay industry was expanding more or less as metal mining was declining many engines were available near at hand. The presence of skilled men to re-erect them and others to operate them would also influence the early clayworks owners. Within a few miles of St Austell there were more examples of nineteenth-century engines than in any other comparable area. Many of these lasted until the period between the wars, when an interest in early engineering was slowly awakening and indoor photography was not too difficult.

Washing clay seemed mundane and less romantic than mining tin with the result that this great collection of nineteenth-century engines went almost unnoticed, even by people who claimed to be interested. The Newcomen Society made only one hurried visit in 1932. Two people could be credited with taking some action. Captain R. S. Alston, who had retired from the Royal Navy, spent a week or two each summer taking photographs. W. K. Andrew worked in the sales department of one of the large clay companies and took photographs of engines that he knew his employers would have to replace before long. He lived to see one engine preserved, but unfortunately not to see it operated by compressed air purely as a show piece for visitors.

The writer likens his visits to being there at the eleventh hour, compared to the industrial archaeologists who have arrived at twelve-thirty. Even at the eleventh hour, which in this case was 1935, there was enough left to give a clear picture of what had been used by past generations of Cornish mining engineers, either locally or at the many places where they worked.

2 Before the Days of Steam Power

When people show an interest in early machinery they usually think in terms of steam engines, but considering the long history of non-ferrous mining the use of steam power has been no more than a temporary expedient of recent years. Two important mining areas reached their busiest days without the mess and danger of early steam power. In the Harz in Germany and the northern lead mining districts of England the elevated terrain and high rainfall provided ample waterpower.

A few items from pre-steam days will not be out of place and the easiest way to describe such machines would be to copy whole paragraphs and drawings from *De Re Metallica* written by Georgius Agricola before 1550. He described and illustrated machines driven by men, goats, horses, and water. Among the varied duties of these queer machines were winding, pumping, ventilating and numerous processes in the treatment of ore.

The examples illustrated here are more recent, but are of types that could have been used before steam power was available. These show how different types of technology overlap and one form of mechanism did not necessarily get replaced by the next development. Fig 1 shows two men working an ordinary handwinch. What is unusual is that the winch is built almost entirely of wood and is not much of an advance on two forked pieces cut straight from the forest. Other hand-driven mining machines, as distinct from tools, include a ventilating fan in Earby

1 A simple wooden-framed winch, such as would have been erected over a shaft by early miners. Here being used when repairing a pump at Greensplat Clay Works, Cornwall, in 1946. A sliding piece in the top could be used to lock the left-hand handle.

2 The Sam Oon Stone on Greenhow Hill, in Yorkshire. It is thought to have been used for crushing lead ore, probably in monastic times.

3 This drawing from Louis Simonin's book *Mines and Miners* shows such similarity to the Sam Oon Stone that it cannot be ignored. This device was used in Mexico, where Catholic influence could possibly have adapted a monastic idea.

Mines Museum and a two man-power ore crusher in the Craven Museum at Skipton. The Sam Oon Stone on Greenhow Hill can hardly be classed as machinery, but if it is studied with the drawing from *Mines and Miners* (1868) by Louis Simonin one can visualize it as part of a primitive machine.

Where more power was required a horse was a natural choice in days when they were widely used for transport. The best known horse-powered machine is the winding arrangement known as a whim in Cornwall and in other places as a gin. It is one of the few machines that show little difference whether used at a mine or colliery. Several old pictures of them exist and at least two full-size ones have been preserved as museum exhibits. Fig 4 shows a typical horse gin. As the horse walked round one bucket came up the shaft as the other went down. When the weight of the descending rope or chain overcame the ascending one the horse in its pair of shafts could brake the drum in a similar manner to going downhill with a cart. Agricola showed these

4 A horse whim or gin, at a colliery near Eckington, south-east of Sheffield, and photographed in the 1930s. For very deep pits several horses were used at a time, but the size of a gin was limited by the maximum length of timber available for the cross bar at the top. The headgear had to be low so that the rope would coil onto the drum. This example worked until the pit closed about 1943, and was the sole means of raising both men and coal. *(M. E. Smith)*.

5 An American improvement on the horse whim. As there was only a brake handle the horse would have had to turn round at the end of the wind. Other similar machines incorporated a reverse gear.

THE "DAVIS" SAFETY BRAKE
HORSE-POWER HOISTING WHIM

(IN POSITION FOR OPERATION)

6 The roller and track of an old horse-driven ore crusher at the Odin Lead Mine, near Castleton in Derbyshire, built in 1823 at a cost of £40. Other examples used a stone wheel without an iron tyre on a stone-paved track.

7 This drawing, also from Louise Simonin, shows a Mexican mill. Although not purely for crushing, for a chemical process was also involved here, it shows that the operator had to step back each time the mule went past. An unexplained feature of animal-driven machines is that they always revolved counter-clockwise, except in Europe or North America. This observation does not apply to reversible winding gears.

machines with the driver riding round on a seat behind the horse, but English drawings usually show a boy leading the horse. Not only Agricola but later German writers and William Pryce in his *Mineralogia Cornubiensis* (1778) showed these winding gears inside a building with a conical roof, yet more recent drawings and even photographs show them in the open air. An almost identical machine, but with a smaller drum, was used for lifting heavy parts of pumps and was more often called a crab. When a heavy part had to be lowered down it was sometimes arranged for the horse to tow a sledge as a brake. Used for the same purpose were capstans where men walked behind four long arms.

However well-trained a horse might be it could make a false movement, so the primitive horse whim was improved in USA so that the banksman could drive it more like an engine and the horse did not need leading. In the Smithsonian Museum there is a horse driven well-boring outfit complete with gear for withdrawing the tools.

Horse-operated ore crushers were used in Derbyshire. These had a wide, flat, iron, circular track on which an iron tyred stone roller was pulled round. Beyond what is obvious little is known about their mode of operation. Was the track loaded and then the roller taken round two or three times (which could be looked upon as a batch process)? Or was there some means of loading and scraping off as a continual process? Just as we have seen with the Sam Oon Stone, Louis Simonin has provided a picture with more than a fanciful resemblance, (Figs 6 and 7).

Even horses become tired and at a time when animals were often treated cruelly several would have to work in relays if much winding had to be done. Winding could include drawing water in leather bags or buckets. On the other hand waterwheels do not become tired, so they had been adapted for winding at an early date. Agricola showed one with two rows of buckets so that it could rotate either way. A man in an elevated cabin worked two handles which were coupled to the sluice gates. This seems to be the first recorded case of a man working as an engine driver, as distinct from an attendant who sets something working and then turns it off when it has completed its work, such as crushing or pumping.

The use of reversible waterwheels spread to England, but no drawings or photographs seem to exist. There used to be a large wheel at Cowshill in Weardale, but the writer found no more than an empty house in 1948 and today the site has been levelled over. One of John Smeaton's drawings shows a waterwheel and reverse gears for winding. These were widely used in collieries, especially in North-East England.

A different application of water power for a mining purpose could be seen at Penrhyn Quarries in North Wales. Although the main part of the quarry was deep it was high enough for it to be drained by gravitation through an adit. This made it possible to run water into the quarry and use a water balance to lift the slates from the different levels. The hoisting was not done up the sides of the quarry but in vertical shafts entered by short tunnels from the levels or floors in the quarry. These shafts had a cage and a water tank coupled together by a rope or ropes that passed over a sheave at the top. Pouring water into the tank when at the surface caused it to descend and lift up the cage with a load of slates.

Waterwheels for pumping were still employed long after their use for winding and even today people still visit one at Laxey in the Isle of Man. All the pump-driving gear has long since disappeared, but it consisted of a crank and rather more than 600ft of wooden rods. These so-called flat rods were supported on wheels, which in turn were on rails. The whole assemblage of

8 Sedling Mine in Weardale. All that can be shown of a water-powered mine in England. The near foundation was for a waterwheel that drove a pump. The house with a hipped roof contained a large reversing waterwheel and drums. Note the small cage abandoned at the right-hand side of the shaft. This photograph was taken in 1948, but the buildings were all demolished and the site levelled some years afterwards.

9 The waterwheel at the Laxey Mines, Isle of Man, showing the L-bob and linkage to the balance weight at the right and to the pump at the left. This has been preserved and can still be seen.

10 Laxey Mine. Flat rods from the wheel to the mine shaft, showing the support wheels working to and fro on a masonry viaduct. This was working for show in 1939 when this picture was taken, but had all gone when the wheel was restored by the IOM Government.

11 Laxey Mine. Inverted T-bob at the mine shaft, with the flat rod at the top. There was formerly a pump rod down the shaft at the far side and a balance weight at the near side. Working for show in 1939, but overgrown and inaccessible in 1974.

12 An early German system of coupling a waterwheel to pumps some distance away. This arrangement was shown by Löhneis in 1617, although no similar system is shown in Agricola's *De Re Metallica* published sixty years earlier.

rods and wheels worked to and fro with a 10ft stroke. The end of the rods away from the wheel was connected to a bob like an inverted T. From one side of this were over 1,500ft of rods down the mine shaft driving a number of pumps, each one delivering to one above. To partly counterbalance this long length of rods there were a number of balance bobs, one of which is at the side of the wheel. Many visitors must consider this to be a unusual arrangement and yet the only unusual feature is its large size.

The Laxey Mine closed in 1929 and the famous wheel ceased to do any useful work, but even in 1946 it was still possible to see waterwheels driving pumps in the china clay works of Cornwall. When the use of flat rods for transmitting power over fairly long distances was first used is not recorded, but Löhneis showed them in double lines with rocking arms between them instead of supporting rollers in 1617.[1] The system used in Cornwall in more recent times was different to either the Laxey or early German flat rods. The flat rods were in the form of round iron bars with forged eyes at the ends. The rods lay in grooved wheels that revolved in the tops of forked posts.

At Wheal Martyn near St Austell a 35ft diameter waterwheel drove a remarkable layout of rods. Following them away from the wheel one soon found difficulty, for the rods passed through the roof of a large clay-drying shed. Making a detour the observer saw a smaller waterwheel and the rods from the large wheel close at hand. A footpath, presumably used by the man who oiled the wheels in their forked posts, helped one to follow the rods through a thicket of prickly bushes to another hazard. The rods worked in a low tunnel through a mass of made ground. Unlike the arrangement at Laxey these rods drove round a corner by means of an angle bob. This angle bob was similar to the one shown in Fig 14, but was not easily photographed. At

---

1 G. E. Lohneis, *Gründlicher und Aüsfuhrlicher Bericht*, (1617, Zellerfeld)

13 A 35ft diameter waterwheel at Wheel Martyn Clay Works, near St Austell, in 1935. It is thought to date from the 1880s when the pit changed ownership. In this case the balance weight was arranged differently to the one at Laxey, but served the same purpose. The first length of rod from the crank was supported on two mine-truck wheels that ran on two short rails. This wheel was restored in 1975 as part of the Wheal Martyn museum scheme.

14 An angle bob for turning flat rods through a greater angle than the single-arm fend-off bob shown in Fig 19. This one, at Wheal Remfry Clay Works, is like the one used at Wheal Martyn. In the foreground there is a thin cable for stopping and starting the waterwheel. This cable also has a small bob to lead it through an angle where one would expect a pulley. Photograph taken by the late W. K. Andrew about 1930.

15 Wheal Martyn in 1939. V-bob leading the flat rods down into the pit. This shows one of the curved supports fastened under the rods to allow for the end rod moving through a small arc.

16 Wheal Martyn. Another bob leading the rods down the shaft. This ran night and day without attention, keeping a disused pit from flooding until such time as it was required. If the water in the sump ran low a self-acting device ran some water back. A lever which was part of this can be seen across the top of the shaft. Note the crude ladders to enable the wheels, etc to be oiled. There is even one on the rocking bob, and although they were often seen they were probably illegal.

17 Wheal Martyn. An 18ft waterwheel also driving a pump, with the balance box at the left-hand side. The photograph is pre-World War II, but this wheel worked for about twenty years after the larger one. Restored in 1973 as part of the Wheal Martyn museum. The spokes are flat iron, but the hubs have deep slots as if intended for wooden spokes. The flat rods from the larger wheel can be seen at the top of the picture.

this point a line of disused rods led to a pump and there was means for coupling them together. At that time interest was more in working machinery rather than relics, so one was led to see where the working rods went to and the next item was a V-bob (Fig 15) that changed the direction of the rods down a steep slope, and finally another bob (Fig 16) coupled the rods to the pump about a quarter of a mile from the waterwheel. Near to the angle bob the forked posts were wider so as to allow side play for the grooved wheels. The last wheel was on an axle between two posts. Just as provision for side play was incorporated, so was there provision for a slight up and down movement near the V-bobs. This was in the form of a long wooden block bolted to the underside of the rods where they rested on a supporting wheel. The lower side of the wood was made to a suitable curve. All the joints in the rods were loose pins, but there was no backlash or lost motion because everything was in tension, yet there was a strange feature almost contraditory to this. If one stood near the rods at the wheel-end they seemed to reciprocate with a steady movement, like the connecting rod of an engine, but at the far end away from the wheel there was a definite pause at each end of the strokes. Was it caused by stretch and contraction in the rods? A heavy balance box near the wheel ensured that the load on the wheel was roughly the same, whether the rods were moving away from the wheel or being pulled back. These two opposing forces were not applied directly to the crank pin, but to a block which held the brass

18 One of two waterwheels at Carloggas Clay Works, near St Austell, in 1935. In this case the first length of rod was supported on a rocking post as at Laxey. Lines of rods entirely supported on rocking posts have been used elsewhere. The balance box is at the far side.

19 Two lines of flat rods at Carloggas in 1938. These rods were led round a small angle by single-arm fend-off bobs, whose outer ends had a small wheel on a curved rail.

20 One of a pair of 42ft diameter, tangent spoke, over-shot waterwheels at Bad Reichenhall, Germany. Each drives three pumps. Built in 1850 and still at work. The object in the foreground is a small fountain, not a helical gear on its side!

21 A small angle-bob at the Karl Theordore brine well at Bad Reichenhall. It is driven by an undershot waterwheel and works in the same way as those formerly at Laxey and Wheal Martyn. It is still working.

bearings that ran on the crankpin. This would ensure that the load on the crankpin was the real load and not a shearing action between the two pulls. The same arrangement was used on all the other waterwheel-driven pumps that were working and can still be seen at Laxey.

The Wheal Martyn pumping shaft caved in during World War II and the whole layout fell into decay and became overgrown, until early in 1975 when the wheel was scheduled as part of an open air museum.

Not many years after Wheal Martyn finished working the only other set of flat rods, at Carloggas near St Austell, fell into disuse, but although it was the last flat rod system to work in Britain a similar system remained in use in Germany. Two waterwheels each driving two lines of rods were seen working at Bad Kreuznach in 1965. The rods were not on wheels but on inverted pendulums and looked to be of recent construction. Other waterwheels driving pumps can be seen at the salt springs in the Bavarian town of Bad Reichenhall.

22 A waterwheel that worked a mine pump at Rookhope, County Durham. In this case the mine shaft was at the side as at Sedling Mine (Fig 8). The wheel had been made at Aberystwyth, so had probably originally worked at a Welsh mine. It was derelict in 1948 when this photograph was taken.

Water power has also been applied to mine pumps in two other ways. In mountainous places where a large fall could be obtained water-pressure engines were used. These had no revolving parts and in some cases were directly coupled to pumps, while others had wooden rods in the mine shaft. Usually placed underground the one known surviving engine in Britian can only be examined by mine explorers.

The other device had no technical name, but went by the name of flop-jack or bucket engine, and obtained its movement by alternate filling and emptying of a box that could rock over in such a way as to control the driving water and drive a pump as well. For many years there has been a mysterious piece of mining equipment at Wanlockhead in Scotland. It looked a bit like a Newcomen engine with the beam mounted on a stone plinth. Recent research has shown that it was worked by a filling and emptying water box that took the place of a cylinder and piston in a conventional engine. A working model is now on display in a nearby museum. A huge engine of this type is recorded at the Ecton Copper Mine in Staffordshire, which pumped from 550ft and must have had a box of at least 1,100 gallons capacity.

For crushing ores stamps have been used from early times, for Agricola described them in such a manner as to give the impression that they were a well-known machine in his day. The essential parts of a set of stamps are a cam barrel and a number of lifters that are raised by cams and allowed to fall vertically, not like hammers. In Cornwall stamps usually had four lifters with iron shoes called heads. These worked in a mortar box and fell in the order 1, 3, 2, 4 five times for each revolution of the cam barrel. A waterwheel could drive up to four sets of four on one cam barrel and besides the water to drive the wheel a supply was needed in the mortar boxes. Ore was fed into the back of the boxes and the pulp discharged at the front and ends by splash in earlier examples, but stamps that existed in more recent times had punched-hole screens to give a more uniform product.

There are records of stamps being used in the lead mines in the north of England, but long before mining declined stamps had given place to water-driven crushing rollers. One set of rollers survived long after the waterwheel was broken up (Fig 24).

China stone is a hard white rock often flecked with purple fluorspar. It is used in the preparation of pottery glazes because it fuses at a suitable temperature. At Tregargus near the Cornish village of St Stephens in Brannel there were several china stone quarries and their associated grinding mills. Together they formed a strange combination of industry and sylvan beauty, for the whole was clothed in luxuriant vegetation, including many wild strawberry plants.

As the mills were driven by waterwheels a short description will not be out of place. Most of them had a waterwheel in the middle of the house or between two separate buildings. In either case two grinding pans were on each side of the waterwheel. These pans were not like an old type mortar mill, nor were they like a flour mill with a fixed bottom and revolving top stone. The pan bottoms were made of china stone pieces called paviours fitted together to form a ridged base.

23 A waterwheel and stamps for crushing tin ore near Lanivet, south of Bodmin in Cornwall. The date of the picture is unknown, but is probably between the wars.

24 Remains of a waterwheel-driven double set of ore crushing rolls at Old Providence Mine, near Kettlewell, Yorkshire, in 1943. One set of rolls was taken to the Earby Mines Museum in 1971.

Above were four blocks of china stone a foot or more cube, which were driven round by four arms bolted to a central driving shaft. The pans had an outer wall of masonry encircled by iron bands and an inner wall round the driving shaft. The construction was watertight as the grinding was done wet. Pieces of china stone were fed in and the final product was a white creamy mud that looked like china clay to the casual observer. The grinding was an abrasion process and the wear on the runners and paviours added to the product, instead of contaminating it. The upright shafts that carried the four arms for propelling the runners were driven by bevel wheels, usually under the floor, and proportioned to step up the speed about three times that of the waterwheel. There was little sound in the Tregargus valley in the evenings after the quarry had finished working, only the birds singing and water splashing with a faint tinkling of gears, for the mills were left working for long periods without attention. One mill has been saved as a future show piece as part of the Wheal Martyn clay works museum scheme.

The same type of mill was used in the Staffordshire Potteries for grinding calcined flint, using paviours and runners of chert that came from undeground quarries at Bakewell in Derbyshire or far away at Fremington is Swaledale. One of these mills is a show piece near Leek. It runs but does not grind.

25 One of the china-stone grinding mills in the Tregargus Valley, near St. Stephens in Brannel, in 1946. The tram lines led to the quarry. On the left-hand side of the archway is a handle for opening the spill valve to stop the mill.

26 The restored flint-grinding mills at Cheddleton, near Leek in Staffordshire. The far wheel has an X-section axle, a feature often seen between the hubs, but rarely between the hubs and bearings.

3 Primitive Steam Power

The history of the development of Thomas Newcomen's steam engine, which generated power by atmospheric pressure acting on a piston after steam had been condensed in a cylinder, is well documented.

In spite of atmospheric engines having a very low efficiency they found great favour in colliery districts where plenty of waste coal could be used. History books suggest that James Watt's improved engine rapidly replaced the atmospheric engine, but this was only true in

27 A Newcomen engine at Elsecar Colliery, near Barnsley, Yorkshire. This photograph, taken in 1934, shows the engine more or less as it was when it was in use. The engine is still there, and although some tidying up has been done since it is the only Newcomen engine outside a museum and on its original site. It dates from 1795 when it had a 42in cylinder. A new 48in cylinder was installed in 1801 and the original wooden beam was replaced by cast iron in 1836. It worked continuously until 1923 and in emergencies until 1930. This engine has parallel motion on the pump-end of the beam, unusual for mine engines, although common on waterworks engines.

28 A Newcomen winding engine at Nibland Colliery, near Mansfield in Nottinghamshire. This is probably one of the earliest known photographs of a steam engine, for it was reputably taken in 1855 and shows Hezekiah Cheetham, the banksman and Mr Hobson, the engineer. Probably taken after the pit closed, for the flat rope (or chain) has been taken off the narrow winding reel. Note the 'beehive' boiler, the heavy connecting rod at the crank-end of the beam to equalise the up and down strokes of the single-acting engine, and the X-section of the drum shaft, similar to the waterwheel axle in Fig 26. Instead of the usual arch-head at the cylinder-end of the beam there is a guide system for the piston rod; no doubt these open-topped cylinders did not need the elaborate parallel motion used for more advanced types of engines.

places like Cornwall where coal was expensive. Even after Watt's patent had expired in 1800 and there were no fees to pay, due to their very simple safe and cheap construction atmospheric engines were still being built. These later engines invariably had a simple form of 'pickle-pot' condenser instead of condensing the steam in the cylinder, or even a Watt-type air pump and condenser while still retaining the open-topped cylinder without a steam jacket. It is known that in the north, especially in Durham and Northumberland, atmospheric engines were used to pump water onto waterwheels for winding, but it is often not realised that many atmospheric winding engines complete with crank and flywheel were built. In the early 1930s one could meet men in South Yorkshire who had worked with atmospheric pumping engines, although they rarely called them such, but usually: 'an old open top engine'.

Fig 28 shows a typical atmospheric winding engine built for colliery use. Many hundreds of these engines were employed in the late eighteenth and early nineteenth centuries.

There is no shortage of literature explaining how Watt improved this simple atmospheric type of engine to produce an engine with three valves. One let steam into the top of the cylinder, the second let steam pass from the top to the bottom (ie to the other side) of the piston. The third valve let steam from beneath the piston pass into a condenser. The first and third valves worked more or less in unison, and it was the condensing of the steam in a separate condenser, not in the cylinder, that made the engine a success. It was this comparatively simple straightforward type of engine that was improved in the South-West and became known as the Cornish Engine.

4 The Cornish Pumping Engine

No excuses are offered for a large part of this book being devoted to this type of engine. In no field of power generation was so much improvement brought about in less than a normal person's working lifetime. An improvement that was not surpassed for many years.

Little was written about these engines at the time they were developed. The only contemporary book on the subject was by William Pole in 1844, but a great deal of it is taken up by mathematics with only one drawing of an engine. Not many drawings were shown in contemporary text books, and the only book to give details of Cornish engines was not published until 120 years after the death of the author. This was John Farey's *A Treatise on the Steam Engine, Historical, Practical and Descriptive.* The first volume, covering the period up to the end of the Watt patent in 1800, was published in 1827, but the second volume, covering the development of the high-pressure Cornish engines, was about to appear when the author died in 1851. When it was published it was without the plates which had disappeared.

29 Model of a Cornish pumping engine in Holman Brothers' museum at Camborne. Although built to a small scale it shows the essential features.

30 Beam of an 1862 engine without a central rib. The nose pins are fixed by four keys, although other engines of similar date had pins fitted into bored holes, or else floating pins. A beam was always a bob, supported on a bob wall and the outdoor platform was a bobplat. But always a beam engine, never a bob engine. This 50in engine originally worked at Tremenheer Mine, near Helston in Cornwall. In 1874 it was sold to the Aire & Calder Navigation and used at Winterset near Wakefield, when it worked a large diameter low-lift pump to lift water from a small to a larger reservoir. Photograph taken in 1951, and although the engine was still there in 1957 it was probably scrapped shortly afterwards.

In more recent times, when Cornish engines have become a subject of historical interest, much has been written. Ten years ago a writer filled a book of over 250 pages of interesting information about these engines, without explaining how they were put together or how they worked.

It must be realised that in essence the Cornish beam engine was a sophisticated version of Watt's single-acting pumping engine. All the essential features of Watt's engine were present: the separate condenser exhausted by an air-pump, the three steam valves, the steam jacket which kept the cylinder hot, the cutting off of the steam supply after part of the stroke and the expansion of the steam for the rest of the stroke. But as Watt only used a pressure of about 5 lb/in² the effect of expansion was very slight. It was left to the Cornish engineers to use higher steam pressures and refine the valve gear to produce the highly efficient Cornish pumping engine.

To explain the general arrangement of one of these engines a model in Holman Brothers' museum at Camborne will be of help. 1 is the beam, although more often called a bob. These were made of two castings and the inside faces gave the appearance of being made in an open mould. In many cases that part of the bob that was inside the engine house was longer than the outside part. In other words the piston made a longer stroke than the pump. The bob rocked on a gudgeon which was supported by two open-topped stools. 2 is the cylinder, in most cases made with a steam jacket. Early engines had the cylinder fitted into the jacket with a rust joint at the top and a flange at the bottom, making three flanges as there was one on the base. Later engines had two joints at the top: one where the cover was bolted on and one where the cylinder was permanently fastened into the jacket. Whatever type of construction was used the cylinder and steam jacket was encased in some form of heat retaining covering. This varied from brickwork to polished wooden staves. 3 is the mechanism for working the valves, called the working gear. 4 is the pump rod of timber with wrought-iron straps and bolts. The water was forced up by the weight of the rod as it descended, but as its weight was more than was needed the excess weight was relieved by the balance bob and box 5. The exhaust steam passed to the condenser submerged in a cistern 6. Some of the warm condensate was pumped to the boiler or boilers in the house 7.

Before studying the working of the engine more consideration must be given to the detailed construction, for an engine built in Cornwall was quite different from a pumping engine built elsewhere, even if working on the same cycle. The pins in the ends of the bobs were either staked in as shown in Fig 30, or fitted into bored holes. It is reasonable to believe that earlier engines had the pins staked in, but there is evidence that this was also done to make a cheaper engine. Both these methods could be found on engines built outside Cornwall, but a feature peculiar to Cornish builders was to have the bearing not on the top half of the nose pin, but to put bearing brasses into polygonal holes in the bob noses and allow the pins to float. This was said to even out the load onto each side, but at the same time it would ensure better lubrication. Another less obvious point is that the pins could be removed, as is shown on old photographs of bobs being moved from one place to another. If one looks at the outdoor end of the bob on the model it is possible to see the brass, or what some might term a half brass.

31 The Aire & Calder Navigation engine, showing an open-top bearing stool and half brass with an end. No holding-down bolts, but the stool is securely wedged into the base that sits on the bob wall.

Owing to the structure of a full-size engine one cannot view it as a whole, so a view inside the model will help (Fig 32). On this view is shown the brass in the end of the bob. Between the top of the piston rod and the end of the bob are two links. On engines built out of Cornwall these were more often than not in the form of a loop, but on a Cornish engine where they were between the two sides of the bob they had to be made with open ends with gibs and cotters. Paradoxically they were always called loops. Almost invariably engines built outside Cornwall had the loops outside the bob (eg Fig 83), while Cornish-built engines, with only a few exceptions, had the loops between the two sides of the bob.

Some engines had a forked cap at the top of the piston rod and a single loop. The cap and piston rod were constrained to work in a straight line by the parallel motion **8**. This parallel motion was usually different to that seen on beam engines in other places in that the thrust was taken on pins behind the piston rod. The thrust pins were attached to a girder that fitted into the side walls of the house. On the model it is a round cast-iron girder **9** although on many engines it was of wood. In many ruined engines houses one can see a bricked-up opening where the girder was put in.

32 A closer view of the Cornish engine model in Holman Brothers' museum

10 and **11** are valve chests know as top and bottom nozzles. The place where the cylinder and working gear are to be seen was always know as the bottom chamber and it was there that the engine man spent most of his time. Invariably there was a comfortable chair and often a clock. A visitor in 1914 reported finding a seconds pendulum in an engine house, presumably to find out how fast the engine was working. Even though known as the bottom chamber there was a space under the floor where there were weights that played their part in the working of the valves, and also one or (more usually) two pieces of mechanism called cataracts. This space could be entered by a trap door, but was an awkward place to inspect because the exhaust pipe passed through it, too high to climb over and too low to crawl under.

From the bottom chamber a doorway led to the boiler house and a flight of steps led up to the floor above. This upstairs part was called the middle chamber, although it was not always a chamber for in many engine houses the floor extended about three-quarters of the distance back to the front. It can be seen on the model that the floor ends by the top nozzle. From the middle chamber more steps led to the top chamber, but not always direct. Sometimes a narrow balcony ran from side to side on the wall just below the stools. In many cases the stairs and balcony if any, were fitted with good quality banisters. From the top chamber two narrow doors gave access to an outdoor platform called the bobplat. There were two doors because one could not walk round the outside end of the bob.

It will be seen that the boiler is lower than the bottom chamber floor in the engine house. This enabled the hot condensate from the steam jacket to return to the boilers by gravity. The use of steam traps to blow away condensate was entirely against the heat-conserving ideas of early Cornish engineers. Some early engines had a steam jacket in the cylinder base, but this probably did no more than increase the temperature of the exhaust steam and result in more condensing water being used; it was a design that did not persist. Cylinder top covers with a

33 A Cornish engine was always put into a house, not the house built round the engine. Here is seen a Cornish engine being erected to replace an old rotary engine. A runway has been built to enable the cylinder to be pulled into the house. The cylinder can be seen laid on its side. Taken at Dubbers Clay Works, near St Austell, early this century, probably by or for the clay works company.

34 An early Cornish engine with a central rib along the beam. This gave no added strength and was soon discontinued by Cornish engine builders, although used elsewhere. This engine had a forked cap and a single loop. This engine was at Wheal Martha (or Luckett), near Callington, and the photograph was taken in 1938 when the engine was broken up after 70 years of idleness. This 80in engine was erected in 1868 in an extended engine house that had once housed a 50in engine.

steam jacket were also used, but discarded due to the extra joints to make when re-packing the pistons. By the time that metallic piston rings had been developed and cylinder covers could remain in place for years, the interest in extremely high thermal efficiency had waned and the use of jacketed covers did not return.

When an engine worked with a high degree of expansion there was a danger of air leaking into the cylinder towards the end of the stroke and spoiling the vacuum in the condenser. This was avoided by having the piston rod packing in two tiers separated by a lantern brass. A small steam supply was fed to this space to prevent air being drawn in, and on the model the small pipe can be seen. This method was used much later on the low-pressure end of steam turbines.

The Working Cycle

Having studied the layout and construction of a Cornish engine it now remains to describe how it operated. To do this it is more convenient to look at a waterworks engine, because being in a more roomy house it is easier to obtain a view with all that is required for an explanation.

At the beginning of the cycle, with the piston at the top of the cylinder, there appears little sound or noticeable movement. But there is movement for the rod **1** is slowly creeping upwards. Soon it lifts the lever **2** which releases a catch **3** and the bottom arbor (or spindle) and handle **4**

35 A Cornish engine in a waterworks, used here to explain the working of a mine engine. Note that this engine has a single plug rod, with the steam arbor **5** high above the other two arbors (see also Figs 81 and 82). Mine engines often had two plug rods, and all three arbors close together. Taken at Kew in 1937.

make about a quarter turn, the handle moving upwards. This movement is brought about by an underfloor weight, but its purpose is to open the exhaust valve, which is at the bottom between the two fluted pipes. Besides opening the exhaust valve a rod at the right-hand side opens the injection valve on the condenser and a vacuum is formed below the piston. A second or two after this has happed the rod **1** has also pushed up a catch and allowed the steam arbor **5** to make a similar part turn and open the steam valve, which is the middle one at the top, letting steam onto the top of the piston. The piston descends quite fast and this steaming stroke is referred to as 'coming indoors'.

Part way down the stroke the tappets **6**, which are attached to the plug rod and move with the beam, strike the steam horns **7** and close the steam valve. The steam in the cylinder now expands until it cannot depress the piston any further and the engine comes to rest. During the last foot or so of the stroke the tappet **8** (also fixed to the plug rod) strikes the bottom handle **4** and closes both the exhaust and injection valves. This movement of the exhaust arbor also causes the lower of the pair of sectors **9** to release the upper one and allow the equilibrium arbor **10** to make a part-turn in the other direction. The top handle **11** moves downwards as the equilibrium valve opens. This valve is the left-hand one in the photograph and the expanded steam passes through it and down the two fluted pipes to the under side of the piston, which can then make an upwards stroke, referred to as going outdoors. This stroke is made by the weight of the pump at the other end of the bob.

When the piston is nearly at the top of the stroke the tappet **12** returns the handle **11** to the position shown in the photograph. This closes the equilibrium valve and the cushioning effect brings the engine to a standstill. This last movement has also put the top one of the pair of sectors **9** into a position to clear the bottom one ready for another indoor stroke.

It only remains to see why the rod **1** was moving upwards. Previously it was stated that a device known as a cataract was under the floor. The cataract is almost like a small pump, except that in place of an outlet valve it has a small tap. A weight on top of the pump falls at a rate

36 The Aire & Calder Navigation engine showing the cataract. At the right-hand side is the bottom of the plug rod and tappet for resetting the cateract. The thin chain at the left probably extended to the top chamber and could be pulled up to stop the engine.

depending on how fast water can leak out of the little tap. As the weight falls it pushes the rod **1** upwards. At each stroke the engine re-sets the cataract. In the engine shown in Fig 36 the cataract was not under the floor, but was of more refined construction than usual and the builder had put it on view near the steps. Mine engines usually had a second cateract to give a delay to the opening of the equilibrium valve. Early cataracts were made with filling and emptying boxes and on Cornish-cycle engines built out of Cornwall air cateracts were known, and even double-acting cataracts to control both pauses. There is no other method to control the speed of the engine, opening or closing the steam valve by the handle **13** will only lengthen or shorten the stoke, the first doing damage and the second causing the engine to stop due to the sectors **9** not clearing at the end of the indoor stroke. The degree of expansion can be varied by raising or lowering the tappet **6** by means of the hand wheel **14.**

Under the floor a pin and slot arrangement ensures that holding the bottom handle down also holds the steam inlet valve shut. There is one other adjustable point, the little brass nut **15** varies the time between the opening of the exhaust and injection valves and the steam valve. If the condensing water is warm, as in summer, more will be required and one could arrange for more time to improve the vacuum. The writer can call to mind seeing about two dozen Cornish engines working but never saw anyone make this adjustment, and only knew one person who could explain its use. Some might wonder how the plug rod cleared the exhaust pipe under the floor. Engines with a single plug rod had a cross pipe in the exhaust pipe for the plug rod to pass through. This cross pipe was all in one casting with the exhaust pipe.

Some Cornish Engines

The previous description has referred to engines in general. A closer look at one or two examples can now be taken.

The engine in Fig 37 at the Parkandillick clay works was built by Sandys, Vivian & Co at Copperhouse Foundry, Hayle, in 1852. At least that is cast on the bob but no doubt some parts were altered or replaced at sometime, for the engine was already sixty years old when erected at the clay works in 1912, with a new cylinder provided by Bartle's Carn Brea Foundry. It will be seen that the chimney is detached from the house, but in many cases chimneys were built as an integral part of the engine house. This house had not as many windows as some. The shaft was about 240ft deep and the engine pumped clay-bearing water from an open pit some distance behind where the photograph was taken from. The next photograph, Fig 38, is in the bottom chamber and the engine man is operating the working gear until the pump has picked up its load and the condenser starts to pull a good vacuum. Before long the engine would become self-acting and if the steam supply kept to the same pressure it would continue without much attention. The engineman has his left hand on the top handle, which is coupled to the equilibrium valve, and his right hand is on the bottom handle, which is coupled to the exhaust valve. In the position shown the exhaust valve is closed, because one can see the tappet on the plug rod in contact with the handle.

The next photograph, Fig 39, is in the middle chamber and shows the cap at the top of the piston rod. There is no visible means of fixing the cap to the rod, but the top of the cap is bored out larger than the lower part and two C rings fit into a groove in the rod. Many engines had this feature as a safety device, so that if a pump rod broke there was less risk of the piston smashing the cylinder bottom even if the top cover was damaged. The cap has another bit of clever design about it. The pins on which the loops work are integral with the cap, but to make them very much stronger there is a rib, or dog ear, from the top of each pin towards the top of the cap. A half-brass works on the bottom of the pin and a chambered-out block on the top clears the dog

37 *(above)* A 50in diameter cylinder Cornish engine at Parkandillick Clay Works, working in 1938. Now a museum piece as part of the Wheal Martyn scheme.

38 *(left)* Parkandillick engine. The engine man has his right hand on the bottom handle, which is in the down position holding the exhaust valve closed and also the steam valve through a pin and slot device under the floor. His left hand is holding the top handle up and the equilibrium valve shut. The engine is indoors and the long tappets are also holding the steam valve shut.

39 Parkandillick middle chamber. The dog ears on the cap can just be seen. This engine worked for many years with the two rectangular tin cans riding up and down below the loops.

40 Parkandillick engine. Photograph taken off the steps show in the previous view. The round cast-iron girder for the parallel motion anchorage is just visible at the top. Compare this with the photograph of the engine at Grassmoor Colliery, Fig 80.

ear. Many visitors to Cornish engines have missed this point and the builder of an otherwise excellent model told the writer that there were no such things.

Behind the cylinder is the top nozzle and the three valve spindles can be seen. At the right-hand side is the valve by which steam can be turned on or off, and at the other side is the inlet valve, while the middle one is the equilibrium valve. Other engines had the equilibrium valve at one side and no-one in Cornwall ever said that they had heard of another like the above. Yet surprisingly there was an engine by the same builder and about the same date with this odd feature working at Grassmoor Colliery in Derbyshire as late as 1936, although it was in an extremely neglected state. Fig 40 was taken off the stairs leading up to the top chamber and at the top can be seen the round cast-iron girder which supports the parallel motion anchorage pins. This engine finished working in the early 1950s and after a period of neglect it was cleaned up as a museum and eventually made to operate on a low air pressure.

Little purpose would be served in describing other engines in such detail, but a few others are worth mentioning. Some are interesting because they show different mechanical features and others have different houses. It must not be overlooked that what seems an odd feature on engines that have existed within living memory, may well have been common when there were a large number of engines in use.

41 A Cornish engine at Penhale Clay Works in 1935. This engine, besides pumping up a shaft, also worked a pump by means of a wooden rod coupled to the top of the balance bob. A wire cable worked another pump off to the right of the photograph. Like one of the Wheal Martyn scenes this shows more ladders without sides.

42 Penhale. This engine had a single tappet and horn for the steam valve. Note the hardwood pins in the linkage. These worked silently and without wear in the wrought-iron eyes on the rods.

43 Penhale. A close-up view of the linkage to the exhaust valve. Several variations were used.

44 Penhale. Middle chamber view showing the top nozzle. Cylinder cover bolts hidden by a false cover. The levers had all been turned in a lathe, not flat on the sides (compare Figs 40, 47 and 84).

At Penhale Clay Works, near St Austell, a small Cornish engine was working when the writer first visited Cornwall. Outside it was unusual in that it had a square chimney, although it conformed to the old idea of having the chimney at the corner of the house. Not only did the engine pump from a shaft, but it also worked one or two shallow lift pumps by means of a wire cable. The cable was coupled to the top of the king post on the balance bob, so in effect the down stroke of the pump rod pulled the cable. No maker's name could be seen on the engine and it might have been built of bits of different makes, but an interesting point was that all the levers that formed part of the working gear were turned and not flat on the sides. In other words the builder had done as much work as possible in a lathe.

45 Engine at Lower Ninestones Clay Works, working in 1935.

Nearby at Lower Ninestones was another of the very few engine houses with a square chimney. This engine had been built at Bedford Foundry, Tavistock in 1853. The industrial archaeologists have recorded its previous history but here its mechanical details are considered. The working gear was like the two previous examples in its operation (as were all Cornish engines), but it was different in its construction. The other two engines had two plug rods but this engine had a single rod which was sloted. In place of a pair of horns for operating the steam valve there was a single blade that protruded into the slot in the plug rod, and the tappet was in the form of a block in the slot. There was also a difference in the linkage between the bottom handle and the exhaust valve. This engine had a forked cap and a single loop. Like the engine at Penhale it was broken up before World War II and today it is difficult to say exactly where it stood, as the site has been levelled and the pit filled.

46 (above) Lower Ninestones. Bottom chamber view, showing the slotted plug rod and single horn on the steam arbor, which was without a handle. The rod leading out of the top of the picture at the right was part of the rather indirect linkage to the exhaust valve. Like Penhale the inside of the house was plastered to look like large stones in a regular pattern.

47 (left) Lower Ninestones. Middle chamber view showing the forked cap. On this engine the valve-operating levers merely engaged square holes in the valve spindles, but the two previous examples had short links in the spindles to lessen any side thrust. The spherical weight on the stop valve spindle was considered bad practice.

48 Carpalla Clay Works, 1937. The date 1915 on the house suggests that this was the last Cornish engine erected in the clay country, although the engine was over 50 years old then.

49 Carpalla. A simple straightforward design. The polished covers below the valve-spindle glands suggest that some iron clothing is missing. Compare this with Parkandillick and some of the following photographs which show a top nozzle properly clothed. This picture also shows the steps to the top chamber in a different position.

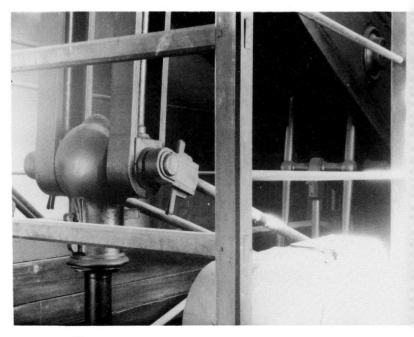

50 Carpalla. Top chamber view showing another type of main cap.

A further example was a 40in engine at Carpalla Clay Works and had been built by Harvey & Co of Hayle in 1864, but the date on the house was 1915. Like other clay-works engines it was second, third, or perhaps fourth hand having worked previously at the neighbouring copper mines. The house was a little more ornate than others and while the engine had no special features it was always kept clean and tidy. It became a war casualty, for when the pit was not working it was allowed to collect water, which softened the rock in the adit and it caved in. After several years of idleness the engine was pulled down and taken away to the Science Museum in London, but even now twenty-five years afterwards it is not on display.

Finally, one view of an engine at Goonbarrow. This was the last Cornish engine to be built, as late as 1914. It was the only new engine in a clay works and was one of very few that was never pulled down and re-used somewhere else.

51 Wheal Remfry. This Engine was like the one at Lower Ninestones in that it had a forked cap and also a slotted plug rod. The top nozzle was cased in, as most engines were intended to be by their builders. In this case there was a balcony across the wall below the bob. Photograph taken in 1936 and the engine was broken up a few years afterwards.

52 Goonbarrow Clay Works. This view, taken before 1934 by an unknown photographer, shows an unusual layout with the chimney near the shaft. This engine, built in 1914, was the last completely-new Cornish pumping engine ever built.

5 Rotary Engines

The engines dealt with so far have all been non-rotative pumping engines, but there were also beam engines with a crank and flywheel. In other parts of the country these would have formerly been called crank engines and more recently rotative engines, but in Cornwall they were known as rotary engines. It is reasonable to believe that at one time they were more numerous than pumping engines, although today far less is known about them. This lack of information is largely due to the fact that this type of engine had become obsolete long before pumping engines. Other reasons are that Cornish engineers never paid as much attention to this type, and also being smaller they never received as much attention from visitors. Sometimes outside view of these engines are seen on old mining photographs, but little would be known of their mechanical details if it were not for the fact that many were used second-hand in the clay works. It made no difference what an engine had done previously, whether winding or stamping, some ingenious clay-works man could adapt it to pump clay slurry.

Cornish rotary beam-engines were entirely different to the beam engines that formerly drove textile mills in the North of England, and even those engines are not properly represented by the objects seen at model exhibitions and labelled 'beam engine'. Engine houses followed much the same lines as those of pumping engines except that they were smaller. In many old houses wooden lintels were used instead of brick arches, giving the windows and doorways a flat

53 A model winding engine or whim in Holman Brothers' museum. The model is based on a late and rather large engine. It should be remembered that many early winding engines wound not cages or skips running on conductors, but egg-shaped kibbles that bumped about in the shafts in an haphazard manner. Note the bell on the roof, it looks as if it could be rung by the engine, but old engineers who remembered them said: 'No – only by hand'.

54 Levant Mine, near Land's End, in 1908. At the left a whim engine of the type referred to as being all-indoors, although the drums were on the outside of the house. This 24in engine was built in 1840 by Harvey and Co, and is now preserved. The pumping engine is complete with a large hand capstan for use when repairing the pumps. In the background is the stamps engine.

55 Higher Bal Shaft at Levant Mine in 1908. Although it looks as if there are no winding ropes they show faintly on the original print. A crank on the end of the drum shaft worked a pump in the mine. The winding drums could be clutched out and the engine used solely for pumping if required.

56 A whim engine at North Roskear Mine, near Camborne. Copied from an old print this shows narrow reels in place of drums. Probably the only visual proof that flat ropes were used in Cornwall. To add confusion, a drum was often called a cage in Cornwall, a relic of the early horse whim where the drum was like a cage.

top. The whole engine was sometimes built inside, but more usually the crankshaft and flywheel were outside. Unlike a pumping engine the stools were bolted down by four long bolts: two inside the house and two outside, with their lower ends nutted or cottered to two iron glands that passed through the wall. Another feature of the houses of rotary engines that differed from those of pumping engines was that the bobplat was often supported at the outer end by a timber horse. Some ruined houses have a single tall window serving both the middle and top chambers. Old photographs show houses without eaves, the roof slates barely overlapping the end walls, and conversely some all-indoor engines had a hipped roof.

It was the general practice in Cornwall for each engine to have its own boilers. In the case of small rotary engines requiring only one boiler, it was often placed in a lean-to. In the case of stamping engines which drove a set of stamps on each side it was the practice to put the boiler house across the back of the engine house, so as not to obstruct the movement of carts bringing ore to the bins feeding the stamps.

Like most pumping engines many rotary engines had the indoor part of the bob longer than the outdoor part. In other words the length of the crank throw was less than half the stroke. The words indoors and outdoors apply just the same if the engine was all-indoors.

Parallel motion was often used, but not always as seems to have been the case with pumping engines. Where it was used it was usual for the plug rod to pass through a loop in the arbor at the bottom of the back links. The reason for this was that the cylinder of a rotary engine was so much smaller in diameter, that there was no need to put an arbor in the bob nearer to the middle for driving the plug rod. It was common for engines to have a forked cap and single loop, as on small pumping engines.

The alternatives to parallel motion were either cast flat slide bars in some form of framework, or round iron ones. These were fixed at their lower ends into lugs on the cylinder, and at their top end to some part of the timber framework that supported the top chamber floor. Where round slide bars were used one could see another Cornish peculiarity, because the sliding parts were made in the form of a stuffing box with two studs and an oval flange. The packing would hold lubrication and tend to dampen any vibration. The same type of construction was used for plug-rod guides.

On many engines the crankshaft was of cast iron and of square section, only being turned in the journals. Often the crank was of such a shape that the outside conformed to the square hole that fitted onto the shaft. Flywheels were built up, and not cast in halves. The usual number of spokes or arms was eight, and the diameter was such that if the flywheel was outside the house its rim sometimes entered a notch in the bob wall. In other cases the wheel rim slightly overlapped the wall. The bearings, which were without sole plates, were bolted to large timbers known as stringers, these in turn being bolted to foundations of stonework generally called loading.

On stamping and crushing engines which had to drive on both sides it was customary to have two flywheels, but due to the difficulty of aligning a shaft in more than two bearings mounted on wood the crankshafts were made in two parts. One part of the shaft and one flywheel were driven from the other by means of a crank arm and a drag link coupled to an extension of the crankpin in the other half.

57 A long-disused whim at Wheal Martha in 1938. Whoever took the photograph managed to be there when the scrapman had smashed one side off the drum and revealed the chain, wound layer upon layer. Another photograph shows that there were three flanges and two chains. Scrapped at the same time as the pumping engine (see Fig 34).

58 Rotary engine at New Halwyn Clay Works, near St Austell. This photograph was taken by the late W. K. Andrew in 1933, just after the engine finished working. At one time it had been a whim at a nearby mine and had been moved and adapted to work two pumps. The pads on the flywheel spokes show where the drum or drums had been bolted on. The little wheel in a forked post would carry a line to a float, or swimmer, in the shaft sump. There was a steam whistle on the boiler house roof. In those days every clay works had its own whistle and there were enough different notes to build a calliope.

Connecting rods were known as sweep rods and were of three types: what might be termed ordinary round wrought-iron rods, thicker in the middle; openwork rods which had two wrought-iron bars forged together at the ends and the intervening space filled with wood, iron blocks or a number of round pins; heavy cast-iron rods of X-section were also used, always on single-acting engines so as to ensure a more even turning movement.

The most interesting point about these rotary engines is the total absence of slide valves and the use of double-beat valves, as on a pumping engine. It may be surprising to learn that many of these engines were like a pumping engine in that they worked on a single-acting cycle. The valve gear was like a pumping engine without the cataracts. This had a slotted link or other device so that the steam and exhaust valves both opened when the equilibrium valve closed, but at the same time it allowed the steam valve to close first. When the exhaust valve closed it allowed the equilibrium to open. This action can still be seen on a preserved engine in Holman Brothers' museum at Camborne. This type of valve gear allowed the engine to run in either direction and in the early days single-acting engines were used for winding. One pictures them as being awkward to stop and start with the nicety required when winding, but in most cases they would drive the drum through a reduction gear.

Double-acting engines were far more suitable for winding and came into use at an early date, with a valve gear not so very much different from the type just described. This had a steam and exhaust valve at the top of the cylinder and likewise at the bottom. The four valve were coupled to two arbors in such a way that one arbor worked the top steam and bottom exhaust and the other arbor worked the top exhaust and bottom steam valves. Weights opened the valves, but a pair of interlocking quadrants allowed only one pair of valves to open at once. A plug rod worked the valve gear, but it will be noticed that this gear had no provision for expansive working. A separate mechanism could be put into operation for expansive working. This was in the form of a valve box where the steam entered the engine and another double-beat valve. A cam with two lobes, either on the crankshaft or on another shaft revolving at the same speed, caused the valve to open for a short period at the commencement of each stroke. A latch, or other holding device, could be used to hold the valve open when required, such as the first one or two revolutions of each wind. In some engines the air pump was driven by the lower end of the plug rod, but more often it was below the outdoor part of the bob, as on a pumping engine.

Engines for continuous running were sometimes built with the valves operated by a revolving camshaft. These had two hand levers, one to lower the camshaft out of contact with the cam followers and another that operated the valves in pairs to get the engine going.

59 Bottom chamber view of the engine at New Halwyn. This was a double-acting engine with a two-arbor gear work. At the left-hand side is the expansion valve with a prop holding it open. The packing has been taken out of the glands suggesting that the engine was probably laid up as a stand-by, before it was finally broken up. The only contemporary description of the working of this type of valve gear is in Julius Weisbach's German text book: *Lehrbuch der Ingenieur and Maschinen-Mechanik* (Brunswick, 1857).

60 A long-disused engine at a clay works at Burngullow, near the present-day White Pyramid. Double-acting with a four-arbor gearwork and air pump indoors. This photograph, taken in 1945, shows the long bolts for holding the bearing stools onto the bob wall, and also how the flywheel rim entered a notch in the wall. A notch at the near side suggested another flywheel, or the provision for one. Replaced by an oil engine and Jackson gear in the early 1920s.

Towards the end of the period during which beam engines were built for winding, ordinary link motion had come into use so that the engine man had better control over the engine, but it should be pointed out that the link motion did not provide for expansive working, the separate valve and cam being retained.

A brief mention should be made of two odd engines, although they might not have seemed odd in the days when a large number of rotary engines were at work, for there could have been others like them. At Hallaze Clay Works there was an engine that had formerly been a crusher engine. It had a camshaft-type of gear, but the piston rod extended up between the sides of the bob and the slides were up in the roof space. Two links, or loops, hung down from a cap-cum-crosspiece to two stub pins in the end of the bob. This construction resulted in a lower building.

At Burngullow clay works there was a double-acting engine with a four-arbor gearwork, so that every valve event could be set seperately. This would seem an ideal arrangement, yet it had a reputation for being a steam eater and had fallen into disuse years before the photograph was taken.

Having studied the working parts of these engines, it now remains to consider how their power was used. Earlier, the horse whim was described and there is documentary evidence that steam power was coupled to similar machines, with the drum on an upright axle. All of James Watt's steam whims were of this type with a vertical rope-drum. They were also fitted with a complex double-spiral balance drum for winding from deep shafts. One old photograph exists that shows one in the far distance, but just discernable as an all-indoors beam engine with the drum outside.

Many years ago, a photograph in *Colliery Engineering* showed a model of a beam engine with the crankshaft and flywheel outside, but geared to a drum on an upright axle. It was an old and unrealistic model, but probably contemporaneous with the full-size engine. These whims had one advantage in that they could wind from different directions, by simply coupling a different rope or chain to the drum. Their weak point was in the use of a right-angle drive in the days of cast gears, and the difficulty of ensuring that the rope would coil properly. Timothy Hackworth tried the same layout for winding up a railway incline and had to abandon it. Considering that drums on a horizontal axle worked by waterwheel had been used for three centuries, it seems all the more strange that this type of steam whim ever existed. There is only little evidence of its counterpart of a waterwheel geared to a drum on an upright shaft.

Far more common was the whim with the drum or drums on a shaft parallel with the crankshaft, either gear-driven or coupled together by some form of flexible coupling. As larger and more powerful engines were built it became the practice to mount the drums directly onto the crankshaft. Any form of single-cylinder engine needed a flywheel, resulting in a type of engine with a dangerously weak feature, viz: a flywheel with its momentum directly at the side of the point where the load was taken. Some engines were built with the flywheel and one side of the drum both keyed to a large bush, which was itself keyed to the crankshaft. This construction went a long way to preventing shafts twisting off between the flywheel and the drum. The model whim in Holman Brothers' museum shows this construction (Fig 53).

If one looks at the illustration of the engine at New Halwyn Clay Works (Fig 58) one can see that the flywheel arms have pads and bolt holes about half way between the boss and rim. This had been a whim at one time and its builders had ensured that there was not going to be a twisting action between the drum and flywheel. It may have had a drum on both sides of the flywheel.

Another old photograph (Fig 56) shows a beam engine, but in place of drums it has narrow reels for flat rope that wound layer upon layer. This had the advantage that when the engine was starting to wind it wound on a small diameter, and as the load decreased the effective diameter increased. This was more common in colliery areas than Cornwall, but was also used for ore mines in the Isle of Man, in the USA, and other distant mining areas.

Any form of winding gear whether horse, water, or steam-powered needs something to show when the load is approaching the surface. In the early days some form of indicator on the chain a few feet above the bucket was all that was required. Perhaps no more than a piece of rag tied on. Later on, when an engine man had taken over from a boy leading a horse in its circular path, something more certain was required. It is claimed that the depth indicator was invented in the Harz mountains in 1773, but it was probably invented in several places as a need for it arose. In Cornwall indicators were known as miniatures, or a set of miniatures. It is not difficult to picture some device in the engine house, or even outside, where a pointer moved up and down a miniature of the mine shaft. The name persisted when indicators had taken the form of a pointer revolving in front of a circular dial. Years ago the writer saw a miniature in the proper sense of the word. It was at Wharncliffe Silkstone Colliery, south of Barnsley, and was little more than a bobbin driven from one of the crankpins. A cord wound onto the bobbin and a pointer travelled up a scale on the wall; just before it reached a mark it caught a bell to indicate that the cage (locally called a chair) was nearly wound up. When the engine was reversed and the cage descended the shaft the pointer did the same on its scale, until the two cages passed each other. At this point all the cord had been paid off the bobbin and the pointer started rising up the scale, indicating the position of the other cage on its upward journey. An indicator of this type can be seen at the small horizontal winder used at the Blists Hill Mine, Ironbridge Gorge Museum in Shropshire.

Stamp mills have been mentioned in the chapter on water-driven machines. Stamp mills driven by steam engines were a common feature of tin mining districts and several old photographs have been reproduced a number of times, but the one shown here is unusual. It has been explained how there were usually four heads in each mortar box, but the one shown has six. The photograph also shows two long links at the side of the sweep rod, which are coupled to a subsidiary bob for driving a pump. In all probability this would be for water used in the ore treatment. It is not known how many heads there were in this set, but it was not unknown for three groups of sixteen to be on each side of an engine.

As explained earlier little would be known about these rotary engines, except what can be gleaned from outside views, if it were not for many of them having been adapted for pumping in clay works. The photograph of New Halwyn Clay Works engine (Fig 58) shows how it was altered. The drum or drums have been replaced and a cast-iron gear fitted to the crankshaft. This gear in turn drives a larger one on another shaft, which has a swape at each end. From the two crankpins two rods transmit the motion via a pair of angle bobs to the pump rods down the shaft. The pipe up which the water was delivered was usually called 'the column' and was between the two rods. This layout was very common and varied from crude to refined, while the power source might be classed as anything that would go!

61 An old stamps engine at Wheal Martha, near Callington in Cornwall, showing the use of timber under the crankshaft bearings.

62 Single Rose Clay Works, near St Austell. A single-acting engine driving a pump in the clay pit and another behind the engine. Like the waterwheel-driven flat rods at Wheal Martyn a pin wheel was used in front at the end of the first rod, but behind there was a rocking post. For some unknown reason the sweep rod was in two parts cottered together. The engine was working in 1935 when the photograph was taken, but finished soon afterwards.

Today's travellers on the road from St Austell to Newquay pass a public house called the White Pyramid, named from the white conical sand tip that used to be behind it. Before the White Pyramid was built and the sand tip was still gaining height, anyone who stood at the roadside became aware of an engine working, for the steady exhaust beat of an oil engine could be heard. This engine drove through a belt and a double-reduction gear to a pair of pumps arranged like the set at New Halwyn. Everything seemed to be good-class engineering. The gears were properly machine-cut and the bearings were in a large cast bedplate. Not far away was a similar plant except that the bobs and shaft top were out in the open. Three other sets were known to have existed, but as the time they seemed so modern that there was no point in bothering to take photographs.

When working it was noticeable that the gears made a slightly increasing sound until the cranks passed over the dead centre, when there was a slight knock and all went quiet momentarily. The men in charge always knew these as Jackson gears, but it was not until many years afterwards that it was realized that an old established gear-cutting firm in Manchester went by the name of P. R. Jackson Ltd. Even if the gears were made for the job the oil engines were second-hand, having driven a small munitions factory in World War I.

During World War II a small mine was opened at Castle-an-Dinas, south-west of St Columb Major, and here an oil-engine driven pump was installed. It is believed to have been the last set of reciprocating pumps ever installed in a mine shaft in Britain.

63 A 22in single-acting engine at Rostowrack Clay Works, near St Austell. Although devoid of maker's name or date it is said to have been built by West and Son of St Blazey in 1851 and to have come from a mine at Lockengate. It was working at Rostowrack about 1880 and ceased in the early 1950s, being now preserved at Holman Brothers' museum, Camborne. Originally the boiler house was on the near side, and even when the photograph was taken in 1938 a difference in the stonework could be seen. Through reduction gears it drove a pump to the right of the house, and also through flat rods and angle bobs another pump a considerable distance behind where the photographer was standing. Note the skip tipping sand at the top of the huge pile of waste. It was hauled up by one of the drums shown in Fig 66.

64 Rostowrack Clay Works, 1945. Another view showing the balance bob and, at the left-hand side of the engine, the winding engine and drums shown in Fig 66. Note the guide pulley where the rope comes through the wall.

65 Bottom chamber view at Rostowrack, just like a small non-rotative engine minus the cataracts and with the addition of a link, so that when the bottom handle lifted it released the steam arbor. Typical of pumping engine houses: a clock and a seat 'to rest when they are weary', to quote from the earliest known drawing of a pumping engine.

A Theory

The extensive Grassington Moor lead mines, in Wharfedale, Yorkshire, made extensive use of water power. Although it is known that a 53ft diameter waterwheel was used to wind from three separate shafts, the details of how this was accomplished is not known. Each shaft was a different depth and a different distance from the wheel. If there had been two shafts a reversing wheel could have been used, with two different sized drums. Such an arrangement has been used in collieries that worked two seams. It is known for certain that much of the successful re-organization of the Grassington mines was carried out by John Barratt, who came from St Austell in Cornwall.

Many years after Barratt's day clay works were winding sand up different sand tips by means of one engine. These sand-winding engines were of any type that happened to be available, and even an old paddle-steamer engine was used at one pit. They were arranged to drive a shaft that had wood-faced cone clutches keyed to them. The rope drums were loose on the shaft and there were also fixed wooden brake-blocks. In use, the engine man turned steam on and pulled a long handle that pushed the drum against the revolving clutch. When the miniature showed that the skip was at the top of the sand tip, he pulled the drum away from the clutch and the skip ran back by gravity. When it was required to stop the skip the engine man moved the lever further, so as to bring the drum up against the fixed brake block (Fig 66). Sometimes two men operated two drums with one engine. The writer suggests that Barratt used this loose drum method, and that both he and clay works engineers had copied it off some long-forgotten mining practice in the St Austell area.

66 Two separate winding drums on one shaft at Rostowrack Clay Works. Friction clutches in the middle and brake blocks at the sides. The indicators were known as miniatures, even though no longer winding up and down like a mine shaft. These drums wound sand skips up onto a tip.

6 Cornish Engines Outside Cornwall

Between the two world wars there were several engine men in Cornwall who would tell of Cornish engines at Millom in Cumberland. These were at the famous Hodbarrow Mine that finished working in 1968. The history of the mine and the company is told in *Cumberland Iron* by A. Harris, but the book does not tell much about the Cornish engines. It is uncertain how many Cornish engines worked at the mine during its 113 years of activity, as in the book there is an old photograph of a rotary engine that looks very Cornish, probably because John Barratt was one of the founder of the mine. In 1936 there was a ruined engine house with curious half-round top chamber windows. The engineer said he had broken up the engine in 1922 and thought it was dated about 1860. He also said that there was no shaft in front of the house. Some old photos give the impression that it was a whim without the flywheel. The photograph reproduced as Plate 7b in *Cumberland Iron* shows this engine at the extreme right-hand side and it seems to have a large six-arm bob in front of the engine. Could it have been an engine pumping from a shaft some distance away? All very uncertain.

What is certain is that there were three pumping engines. The oldest one was built by Harvey & Co in 1876, and the next one by Williams' Perran Foundry in 1878. This engine was said to have been intended for an exhibition in India, but was for sale the next year so it probably started working at Hodbarrow that year or soon afterwards. The other engine was also built by Harvey & Co in 1899. These last two were each the last engine of any size built by their respective builders. All three engines were the same size: 70in cylinders by 10ft stroke and 9ft in the pumps. They did not have the same pump arrangement. The oldest worked a single 24in plunger, while both the others worked a pair of 18in plungers. At one time the engines had bucket lifts from a lower level, but the bottom part of the mine had to be abandoned and the engines pumped from 360ft.

The Perran Foundry engine and the 1899 Harvey engine shared six Lancashire boilers, which also steamed the winding engines and a haulage engine, but it had not always been so. At first the Perran Foundry engine pumped from a shaft that had been sunk into the ore and eventually it became mined out. In 1910 a new shaft was sunk beyond the edge of the ore body and a new engine house built. Even in 1936 it was still being told how a gang of men came from Cornwall and pulled the engine down and re-erected it in a remarkably short time. It was probably some of these men who in later years used to tell visitors to Cornish mines about engines at Millom.

Mine engines usually had a cataract to give a pause at the end of the indoor stroke. These engines had the cataracts connected to floats or swimmers in the sump at the bottom of the shaft, and if too much water was not in the mine the engines paused long enough to take a photograph by daylight. Synchronised flash bulbs had not come into use when Fig 70 was taken.

The engines were kept in good condition and worked with a reasonable degree of expansion, but at times had worked under very bad conditions. On several occasions inrushes of sand had threatened the mine, and it was only saved by the engines being able to pump a thick mixture of sand and water. In the end the lower part of the mine containing about 60ft of ore had to be abandoned in February 1935.

The 1876 Harvey engine had worked for days on end with 10 tons of anchor chain hung on the pump rods to make it pump a mixture of sand and water, that the engineer described as being 'as thick as porridge'. In later years it worked only if repairs were being done to one of the other

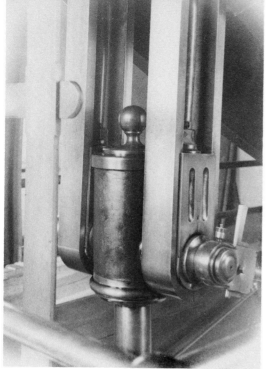

67 *(left)* Hodbarrow Mine, showing the main cap and loops of the 1899 Harvey engine.

68 *(below)* The 1899 Harvey engine at Hodbarrow Mine.

69 *(right)* Hodbarrow Mine, A photograph of the Perran Foundry engine taken by daylight during the indoor pause while the engine was working. The joint where the cylinder was fixed into the steam jacket is just visible near the bucket.

70 *(below)* The Perran Foundry engine at Hodbarrow Mine, Millom.

71 Kathleen Pit at Kennedy Brothers' group of mines near Askam in Furness in 1936. To the right of the balance bob, and laid on its side, is a dipper or bailing device to put in place of the cage if the pumping engine at Violet Pit had to stop for repairs. At the extreme left is Park Mine with a light-coloured engine house.

engines. It worked until the autumn of 1937 and then some timbers in the shaft caved in. A scrap man was called in, but he hesitated too long and at 5.40pm on the following New Year's Day the whole front of the house fell into a crater that formed round the shaft. It was thought that the bob struck the column, for, when the bottom of the shaft was reached from another part of the mine, it was found to be filled with pieces of broken cast iron. Nearly a year after it fell the wreckage was still there, too precarious to be approached. Whether it was eventually salvaged for scrap, or became totally engulfed, is not recorded. The two other engines at Hodbarrow finished working about 1960.

On the other side of the Duddon Estuary in Furness there were two more Cornish engines. One at Kathleen Pit had been built i 1851 at St Austell, but in those days it was spelt St Austle. There seemed little doubt that this engine had been bought second-hand, for a stone shield over

72 Bottom chamber view of the Kathleen Pit engine. Only two other engines built by St Austle Foundry are known to have been photographed and both had double links to the exhaust valve as shown here. The drain pipe from the steam jacket can be seen at the right-hand side.

73 Middle chamber view at Kathleen Pit. The cast-iron plates to fit around the top nozzle were leaning against the wall.

the door bore the inscription K CB & M 1859. K stood for Kennedy the family name and the other letters for Charles Burton and Myles the three partners in the firm. Myles Kennedy had a dredger named after him and it could often be seen at Heysham harbour. This shows that the engine was 8 years old when Kennedy brothers erected it, but that is not the whole story because a small piece of stone had been skilfully let into the bottom of the stone shield with the words 'Removed here 1883'. On the roof was a weather vane in the form of a crossed pick and shovel, looking not unlike the crossed hammers seen in Germany or Austria wherever mining is carried out.

One side of the bob was broken across the main gudgeon, and had been repaired by means of a heavy cast jaw-piece and an odd-sided king post. This king post's short side rested on the jaw piece and the longer side fitted against the unbroken side of the bob. A pair of tie bolts from the king post were anchored round the nose pins. Fortunately for the repairer the nose pins were of the fixed type, otherwise it would have been an even more complicated piece of work. As it was it looked an elaborate method, instead of casting a new side, especially in an iron-making district, but the work was probably done without taking the bob out of the house.

The balance bob, complete with a wooden box of weights, looked very Cornish, except for the liberal use of paint, that looked like the local hematite in colour. A closer look at the balance bob revealed that it had a queen post and tie bolts underneath, where it could give no added strength. The reason for this construction became apparent when looking at an engine in Cornwall. At Goonvean Clay Works there was an engine with such a balance bob, and from the pin at the bottom of the queen post a line of flat rods ran to a china-stone quarry. As the engine performed its normal duty of pumping clay slurry it also took care of any drainage into the quarry. A second visit to Kathleen Pit revealed that the pin at the bottom of the queen post was either turned, or worn smaller in the middle. Whether it had pumped from the nearby Nigel Pit, or if this was a relic from its former use will never be known.

In the middle chamber some cast plates were reared against the wall. They had obviously been intended to fit around the top nozzle. The long one carried the words: 'Hodge Engineer St Austle Foundry'. A cast-iron arch for a doorway is on view in St Austell with the words: 'St Austle Foundry 1849'. The engine did not work much in later years, because the mines were drained by a Hathorn Davey engine that had formerly been at Stank Mine.

On the extreme left of Fig 71 can be seen another engine house built of light-coloured stone. This was Park Mine and the engine was one of the most interesting seen, although in a very neglected state. It was said to have last worked in 1922. The most noticeable feature was the bob which was of a lattice or openwork form, but still retained the Cornish practice of using two castings. In place of curved edges it had flat edges and just a suggestion of a corner above and below the main gudgeon. If one compares it with the drawing of Wicksteed's engine of 1838 at the East London Water Works it is obvious that there is a great similarity. The drawing also shows Wicksteed's engine to have a single plug rod and the steam arbor at the top of tall weigh posts, just like Fig 35 used to explain the working cycle.

Another feature of Wicksteed's engine was that the inlet valve was at the back. On the engine at Park Mine there was also a single plug rod and tall weigh posts but an ordinary top nozzle. A tantalizing point was that this was not the original cylinder for it had the words 'Park Foundry 1877' cast on it, and a bricked-up arch in the side of the house indicated how it had been taken in. It could not have been taken in through the usual large doorway at the rear, because the boilers were across the back. This in itself suggested a steam inlet at the back originally. The engine was arranged for the whole steam supply to pass through the jacket. This explained the former use of two large blanked-off flanges on the engine at Kathleen Pit.

A few words about the mine will not be out of place. The hematite in this area is not in seams like coal, nor in lodes like most ores, but in large irregular deposits that could only be found by

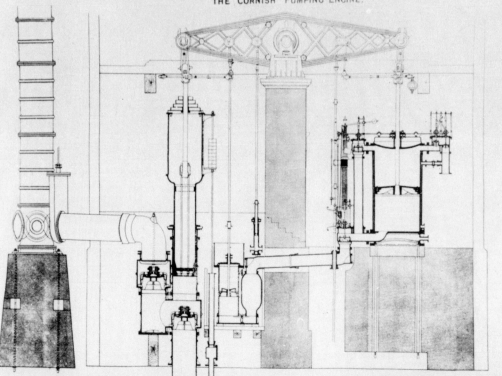

EAST LONDON WATER WORKS

THE "CORNISH" PUMPING ENGINE.

LOT 3.

WHEATLEY KIRK, PRICE & GOULTY.
ELECTRICAL & MECHANICAL AUCTIONEERS.

74 *(opposite, top)* Park Mine in 1936. This 60in engine was built in 1838, probably by Harvey & Co, for the Carlisle Canal and was moved to Park Mine in 1856. It is probably the oldest Cornish pumping engine that has ever been photographed. At least one other local hematite mine was like this, with the winding engine behind the pumping engine. There were indications that the headgear had carried another pair of sheaves at right-angles to those seen and that it had been a four-cage shaft. No other photographs have been found at this mine. The edge of the broken ground is just in view at the bottom of the picture.

75 *(opposite, bottom)* The East London Water Works engine, probably from the sale catalogue in the early 1890s. The engine beam was like the one at Park Mine. Note the steam valve at the back of the cylinder and also the steam-jacketed cylinder base.

76 On the Park Mine engine the main cap was fixed to the piston rod by a gib and cotter. This gave no protection to the cylinder base if the engine had a severe bump, so someone had removed the distance bars from the loops and inserted wood blocks with sloping ends, so that they would be forced out to prevent the momentum of the heavy bob damaging the cylinder base. It was not usual to see loops with Gothic ends instead of radii.

extensive core drilling. Henry Schneider spent a fortune searching for ore and when about to give up found a large deposit. The firm of Schneider, Hanny & Co started Park Mine and the iron works at Barrow and it is no exaggeration to say that Park Mine made Barrow. So much ore was mined that a hole several hundred feet deep and as large as a small town was formed. This was not an excavation, but a subsidence or broken ground.

Henry Schneider had a steam boat on Windermere, which is still in use, although its original engines were broken up soon after World War I. This boat led a lecturer to start doing research about Henry Schneider and he was sent to the writer for photographs of Park Mine and also wanted to know where it was. On being told that the buildings were all blown up as a war-time exercise he was not deterred, but went to the pile of rubble and saw one corner of the stone tablet that carried the words 'S. H. & Co 1856 Park Mines'. Eventually a party, including the lecturer and the writer who had taken the photograph thirty years before, assembled at the ruins one Sunday morning in the spring of 1968 and salvaged the date stone, which is now in Barrow.

These and a few other mines in Furness produced the showy specimens of kidney ore seen in museums and cabinets all over the world. If one sees a new mineralogy book for sale it is a good test to look up kidney ore. If the book says it *is* found in Furness it is 40 years out of date. On the other hand if it says *was* found the book might be worth buying.

77 A Cornish engine at Grassmoor Colliery near Chesterfield in 1936.

78 Bottom chamber view of the engine at Grassmoor Colliery, showing a slightly different construction. The arbors worked in bearings that were not fitted into the front of the weigh posts, but in bearings half-way between the back and front. The bearing top halves were held down by a taper wedge that had an adjusting knob on the front. The lever and catch for controlling the indoor pause have been removed, but the rod from the cataract had been left in position. A strap is holding the bottom handle down while the photograph was being taken.

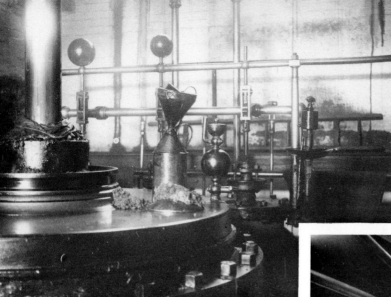

79 Middle chamber view of the Grassmoor Colliery engine, showing the narrow balcony across the wall.

80 Grassmoor Colliery engine. Taken off the balcony this is one of the few photographs showing a round iron girder and the parallel motion anchorage.

Earlier, when describing the engine at Parkandillick, it was mentioned that a similar engine worked at Grassmoor Colliery near Chesterfield. Fig 77 shows it to have had a hipped roof, like all-indoors whim engines. One of the interior views shows the round cast-iron girder for the parallel motion. This is one of very few cases where such a feature is clearly shown, together with two long bolts tying the anchorage pin to the wall, just below the sole plate that the stools stand on. An old drawing shows a similar arrangement with bolts into the back wall. One would think that in making the final adjustments when erecting an engine it would be necessary to rock the top of the girder slightly backwards or forwards. At one time a broken girder lay outside an engine house in Cornwall, and it had a square frame on the end. Whether it could have been twisted in this frame and then rusted in, or if it was all in one with the frame which would have to be moved in the wall and then cemented in is not known.

81 Bottom chamber view of a small Cornish engine at Thryberg Hall Colliery near Rawmarsh in South Yorkshire. This had the perpendicular pipe on the centre line of the engine as was usual with small engines, but the exhaust valve was on the side.

82 Middle chamber view of the Thryberg Colliery engine, with rather plain woodwork. The tappets for the steam arbor can be seen, because they were high up on a single plug rod. The four bars from the cylinder cover were to prevent anyone being banged on the head by the loops. Why were they not required on other engines?

83 A different type of main cap and loops, but this was not a true Cornish engine, having been built in Plymouth. Note also that the loops were outside the beam. The cottered cap and top brasses on the pins add credence to the tradition that it had been a winding engine.

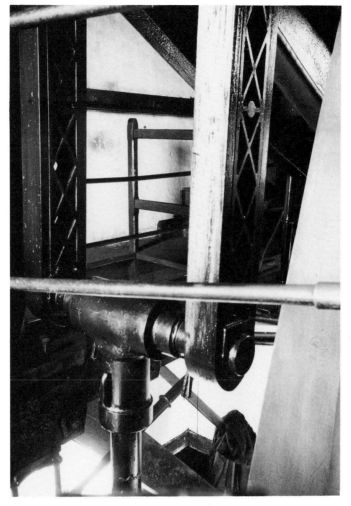

Figs 81, 82 and 83 show what can reasonably be called a Cornish engine for it was built in Plymouth by J. E. Mare, but the date had been chiselled off. It was at Thryberg Hall Colliery near Mexborough in South Yorkshire. A visitor to it in 1914 wrote in his notebook that it had two boilers and that it had had a new cylinder about 14 years before. In later years it was steamed from a battery of boilers, probably via a reducing valve.

This engine had a single plug rod, like some others that have been mentioned, but it also had an arrangement not seen on any other engine. The pipe leading from the top of the cylinder known as the perpendicular pipe was on the centre line, as was usual on small engines, but the plug rod did not pass through a cross passage in the exhaust pipe, because the exhaust valve and pipe were at one side. It was said that the engine had formerly been a winding engine and although the use of single-acting engines for winding has been recorded it is difficult to picture one that had the steam arbor out of reach. J. E. Mare is known to have built both engines and waterwheels, but only one other piece of his works seems to have been photographed. This was a single-acting rotary engine at South Caudledown Clay Works. Neither of these last two engines described had a balance bob, with the result that steam was admitted for the full length of the indoor strokes and the outdoor strokes were made with a minimum opening of the equilibrium valve. Just the way to spoil the economy that Cornish engineers had striven to achieve.

84 A Cornish engine at Preston Grange Colliery near Edinburgh. A simple design without a steam jacket and considerably altered at some period. The photograph shows two separate ports at the top of the cylinder. The levers for working the valves had the fulcrum at the end and the rods from the arbors pushed up, instead of pulling down. A few other engines with this arrangement have been known, but there is not another still in existence. The gland made in two halves might have been a repair, made that way to get it in position without taking the cap off the piston rod.

In 1938 a number of people, some of them Cornishmen and others from away, but interested in the subject, thought that they knew of all the Cornish engines still working. They also knew that an engine had been built early in the present century for the Dorothea Slate Quarry in North Wales and were quite confident that they would have heard about it if it was still there. At this time a disused engine at Wheal Kitty was broken up. This engine was of interest because it had two separate ports at the top of the cylinder, one for the steam inlet valve and another for the equilibrium valve. Some time before a similar double-ported engine had been broken up at Lady Pit near Poynton, without being photographically recorded. And now the last example was thought to have gone.

Nine years afterwards, and during which period there had been the war and a frantic drive for scrap, a strange turn of events took place. A group of model builders in Edinburgh went on a visit to Preston Grange Colliery and eventually this was reported in that weekly mine of information *Model Engineer*, which said that the members saw a Cornish engine working. This caused many questions to be asked, and it turned out to be another double-ported engine. The reason for its existence being unknown, except for those working at the colliery, was that it could not be seen from any public road or the railway. Now it forms the centrepiece of a museum being set up by the local authorities (Fig 84).

A few years later a reader's letter in the same periodical told about narrow gauge locos in Wales and various other items, and then casually ended up by mentioning a pumping engine at Dorothea Quarry. Some photographs were obtained in 1951 just as electric pumps were being installed. The owners even promised to leave some water in the quarry if it was dry weather, so that it could be worked when their visitor from Yorkshire arrived. Since those days many visitors have seen it and even coach parties have had it pointed out to them. Fig 85 shows it at work. This engine did not have a balance bob, but the indoor part of the bob was weighted. Besides the weights the bob was different to the usual practice in that it was built up of plate, either wrought iron or steel. Probably steel, although wrought iron was still being used at that time. A similar construction was used in 1912 for the Goonbarrow clay works engine shown in photo 52. These two engines had much in common, although quite different in size.

85 A Cornish engine working at Dorothea Slate Quarry, near Caernarvon, in 1951. This 68in engine was built in 1904 by Holman Brothers of Camborne. The wall round the top of the shaft has been broken down for the installation of electric pumps. This engine is now preserved. Two winding engines with the drums outside the houses can be seen to the left of the pumping engine.

7 Vertical Winding Engines

We have seen how Cornish pumping engines were used in many places outside Cornwall, due to their reputation for economical and reliable working. One cannot help but have a suspicion that they were used in some cases simply because they were available second-hand. Especially does this apply to collieries, where engines could be seen steaming full stroke without any thought given to economy.

There is little evidence of Cornish winding engines being used away from Cornwall. The reasons are not difficult to see. A Cornish mine produced a comparatively small tonnage of valuable ore, while an iron mine produced a larger amount of less valuable ore. The difference compared to a colliery is even more marked, for in that case the output is not only larger but more bulky. The result was that colliery winding engines were developed with the emphasis on fast winding and quick handling, quite different to the early Cornish plug whim. Even in 1844 it was said that Cornwall was a century ahead of the rest of Britain in pumping, but a century behind in winding. It should be noted that as Cornish mining changed from copper to tin, which was a far less concentrated ore, Cornish engineers found it necessary to increase their winding speeds.

When studying colliery engines one cannot associate different types with different parts of the country and even engines built in Scotland could be seen in Yorkshire. There was one type that had a local preponderance, but not enough to make it more numerous than two-cylinder horizontal engines. This was a type sometimes called a Crowther engine, or more often a Durham winding engine. These engines had the drum at the top of a tall house and the vertical cylinder below. A parallel motion constrained the crosshead and a plug rod from one of the parallel-motion arms formed part of the valve gear, not unlike a Cornish plug whim, except that there was no expansion gear. Being condensing engines they would be as awkward to handle, but with the added fault that they were unbalanced. Extra weight in one half of the flywheel rim no doubt helped to even out the movement. Faults or not, several lasted until after the Nationalization of coal mines. Many were of large size with cylinders up to sixty-eight inches bore by seven feet stroke. A small one has been rebuilt at Beamish Museum, County Durham.

In the mining section of the Science Museum there is a model of the surface layout of a colliery with one of these engines, but it is not in a masonry house but in a wooden framework, which gives the impression that it has been made that way so that visitors could see the interior. Fig 86 shows such an engine in a wooden house, but it is only a small engine. Some of these engines had the inner crankshaft bearing supported on a wall built as part of the house. Others had a massive timber A-frame, while iron pillars were also known.

A few of these north country engines had eccentric-operated valves. One is shown in *Timothy Hackworth and the Locomotive*[1] but it is upside-down and poorly reproduced. It is known that this engine had a balanced slide valve, which was easy to work by hand, and it had a Carmichael-type gear with a single eccentric. Fig 87 shows a similar type engine modelled at three-quarter inch to the foot scale.

The engine at Auckland Park Colliery (Fig 88) shows a feature often associated with Durham winding engines, for at the left can be seen a sheave suggesting that it wound from a second shaft. This sheave carried a chain on which was hung a balance weight. This weight operated very much like the miniature at Wharncliffe Silkstone Colliery, in that it always set off on its gravity-assisted downwards journey when the engine started to wind. As the cages passed in the main shaft the balance weight reversed its movement and so helped in the retardation.

---

1 R. Young, *Timothy Hackworth and the Locomotive*, 1928 (Locomotive Publishing Co)

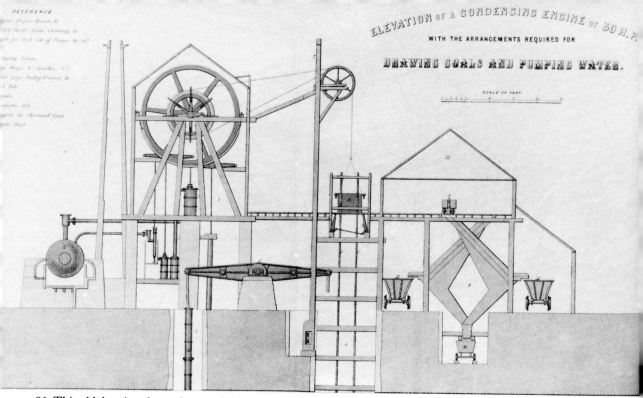

86 This old drawing shows the use of flat ropes and drums with projecting arms, but is more remarkable for the timber-framed engine house. The boiler is the egg-ended, externally-fired type once common at collieries and iron works, and still seen at odd times converted to water tanks.

87 A model of a Durham-type winding engine showing the A-frame and also the parallel motion.

Some balance weights were in the form of a chain with large heavy links, which piled up in a small shaft. This gave the effect of a diminishing weight as less was required and then an increasing one as greater retardation was needed.

This type of engine associated with County Durham seems to have had a small use in other coal mining areas, and examples are known to have existed at Abercarn Colliery in Wales and Lady Pit near Poynton in Cheshire.

Although not a mine engine, one that ought to be mentioned at this point was at Weatherhill near Stanhope. This hauled railway wagons up a long incline from some quarries. After being replaced by a tug engine it stood idle for a number of years and then in 1932 was placed in York Railway Museum.

88 Winding engine at Aukland Park Colliery with a balance chain. Remains of beehive coke ovens are in the foreground. The photograph was taken in 1948, but the site is now levelled.

89 A much larger combined winding and pumping engine of the same type, erected at Ryhope Colliery in 1855 and working until 1933.

90 Erecting the engine at Ryhope Colliery in 1855. There were two engines in separate houses back to back. The capstan, or crab, had a very small diameter drum to lift very heavy loads.

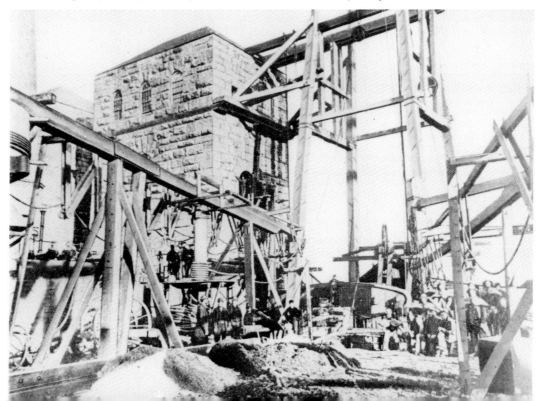

91 Single-cylinder vertical winding engine at Wharnecliffe Silkstone Colliery in 1943. The valves were operated either by hand, or by one of the two rods with tappets. Note the war-time blackout on the windows.

92 Wharnecliffe Silkstone Colliery. Another view on the ground floor.

In the South Yorkshire and neighouring coalfields there were a few engines with the drum at the top, but as far as is known not of the Durham-type with plug-rod valve gear and condenser, and most did not have the parallel motion gear characteristic of the Durham engines.

At Wharncliffe Silkstone Colliery, south of Barnsley, there was a single-cylinder winding engine, built in 1855 by J. & G. Davies of Tipton. Although not a large engine, with a cylinder about 34in by 5ft stoke, it was of interest due to an unusual valve gear. A handle could be moved up and down to operate the four double-beat valves when working the engine by hand. The same handle could also be moved either left or right so as to engage tappets on one of two reciprocating rods, which were coupled to a small double-throw crankshaft, which was driven by a drag link from the end of the main crankpin. The engine drew four tubs from a depth of 460ft. At some time it had been altered, because it had originally had flat ropes winding on two reels and between them a balance rope, as at Auckland Park.

93 Wharnecliffe Silkstone Colliery. Like the large double-cylinder engine at Diamond Pit (Fig 99) this engine also had the valve gear driven from a separate shaft. A steel strap had been shrunk onto the crank. The board on the wall recorded the dates for recapelling the ropes and eventually changing them.

Not far away at Rockingham Colliery there were two twin-cylinder winding engines with the drum at the top. One was replaced by a horizontal engine and the other worked for much longer. The crankshaft bearings were mounted on two cast iron A-frames. It had quite ordinary link-motion, but the interesting feature was that it had Durham-type parallel motion (Fig 94). There was a floor in the house level with the tops of the cylinders and the driver's position was on this floor in the back left-hand corner. The engine drew four 10cwt tubs from 1,080ft. No maker's name nor date could be seen, but it was said that both engines had been built in 1875 by the Lilleshall Company, near Wellington in Shropshire.

What looked like a duplicate of one of these engines was seen at Langwith in North Nottinghamshire. In this case the driver sat between the two cylinder covers and below the

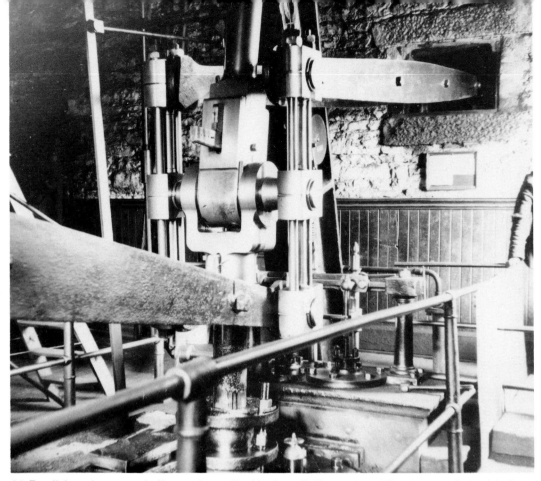

94 Parallel motion on a winding engine at Rockingham Colliery, 1943. The piston-rod gland had two studs and nuts, but was round not oval – a point that many model builders should note.

drum. A large window enabled him to see the top of the shaft and watch the tubs being taken out of the cages. This was a practice that was frowned upon in some places, as it was thought that the driver might ignore the signals and work by eye.

Two photographs, Figs 95 and 96, show the outside of one and the inside of the other of a pair of engines at Wombwell Colliery, near Barnsley. These were built by John Musgrave & Sons of Bolton, probably in 1856 when the colliery was sunk. The inside view shows that instead of parallel motion the engine had slide bars, which formed part of the support for the bearings and also for some of the masonry. The link motion was of the Gooch type and showed signs of alteration. The whole engine had been pulled down and put in a new house so as to wind from another shaft. When the photograph was taken at Easter 1943 there was a man living nearby who had worked on the change-over. The engine was only used for man-winding from 1,800ft.

The other engine, which is only shown by an outside view, wound two tubs of one ton capacity from 300 ft. It wound by means of flat ropes and it was said to have been the only flat-rope winder in use, and that the ropes had to be specially made for it. Ten years afterwards the writer was shown a flat-rope winder at High House Pit near Old Cumnock in Ayrshire and was told *it* was the only one in use and that the ropes had to be specially made for it! Inside the house of this engine at Wombwell there was an unusual piece of carpentry in the form of a flight of spiral stairs, made by morticing the treads into what looked like an old telephone post. It would be inaccurate to call it a staircase, for it was open. A hanging rope served as a hand rail.

95 Wombwell Main Colliery in 1943. The square shape of the engine house is noticeable; this was the engine with flat ropes and the whole engine may have been narrower than an engine with a drum.

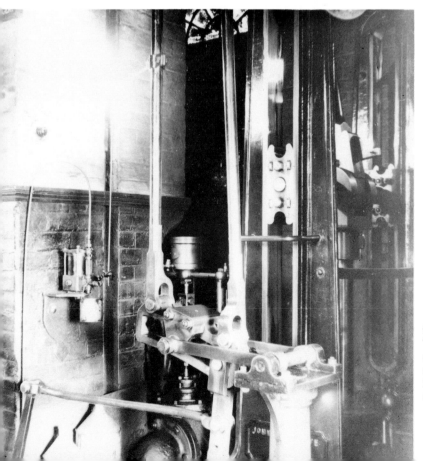

96 Wombwell Main Colliery. The ground floor of the other engine showing the guide bars and Gooch link-motion. Taken from between the two cylinders.

97 Diamond Pit at West Riding Silkstone Colliery in 1943. George Hudson's York and North Midland Railway in the foreground.

When George Hudson built his York and North Midland Railway it formed a junction with the Midland Counties Railway at Altofts just to the north of Normanton, near Wakefield. Whether there was a colliery there then is not certain, but thirty years afterwards there was a flourishing concern in the very apex of the junction. It must have been in a sound position for a new winding engine was required, but there was not much room. Bradley & Craven of Wakefield built an engine for the Diamond Pit in 1866 and it was so placed that only a narrow path separated the engine-house wall from the railway. It was quite a large engine, with two cylinders 40in by 6ft. At some time it had a new drum fitted by Robey & Co of Lincoln, the valve chests had also been replaced and a cut-off gear added. This gear was put into operation when the engine reached a certain speed after about three revolutions, but it was so arranged that it did not put the engine onto non-expansive working as it slowed down. The engine was in very good condition, but even when the photographs were taken in 1943 its days were numbered, for the pit was almost exhausted.

Like some of the others described this one had the bearings mounted on top of two walls, but unlike any of the others the driver's position was at the side of the drum in the top rear left-hand corner of the house. He had just got up off his chair and looked out of the window as photograph 97 was taken. The driver said he had wound for forty years and had 'never been up in the top'. This was his way of explaining that he had never over-wound a cage into the headgear. The engine wound from 1,500ft with, it is believed, four-deck cages.

In the late 1920s Wath Main Colliery near Mexborough, was faced with the problem of putting a new engine into a space obstructed by a railway and a modernised version of this engine was considered, but eventually turned in favour of an engine with the cylinders above the crankshaft.

When Bradley & Craven Ltd published a book on the history of the firm this piece of good-class engine building was never mentioned.[1]

---

1 W. A. Craven, *Bradley and Craven, the first 100 years*, 1963

98 Diamond Pit. Driver's seat at the extreme right-hand of the photograph. Note the dumbell shrunk into the drum.

99 Diamond Pit. This shows how a drag link and a slender crank drove a separate short shaft that carried the eccentrics, duplicated on the other side of the engine.

100 Diamond Pit. The tops of the cylinders were about level with the ground floor. The maker's name and date was on an insignificant brass plate mounted edge upwards on the right-hand guide frame. Like the Wharncliffe and Wombwell engines this had crossheads and guides instead of the parallel motion of the Durham-type winders.

8 A Continuous Winding Scheme

Earlier, when considering Cornish winding engines, it was seen how some early ones had the drum on an upright shaft. This installation was so much like an arrangement near Burnley that one cannot help but wonder if one led to the other. In place of the Cornish-type of rotary beam engine there was what may be called a traditional-type beam engine, with an entablature supported on pillars. Outside, on the end of the crankshaft, a small bevel-wheel drove a larger one on the upright shaft, but in place of a drum there was a grooved wheel. A chain from the wheel went down a drift, and presumably around another grooved wheel, then up the drift to the starting place. The wagons had a notched plate at the top and the chain laid in the notch, so hauling the wagons up and lowering the empties down. Near the top the chain passed over a small wheel, so that the wagons were released and could be manhandled. In a like manner the chain was lowered into the notch as each wagon set off on its return journey.

Unfortunately it was half broken-up when first seen and only one picture was taken (Fig 101). This little relic of the past is believed to have been called Boggart Hole, although it no doubt had some more correct name. It was at the other side of the road near Townley Hall, on the outskirts of Burnley in Lancashire.

There was at least one other drift mine with this system of chain haulage in the area. The same method was widely used for taking coal from collieries to canal wharves, railway sidings or other points where a regular supply was required. One route passed through the centre of the town and in one place went through a narrow opening in a terrace of houses. It was fenced in by a birdcage to prevent any illicit supplies of coal being delivered, which must have been very tantalizing for the householders in their gardens seeing coal passing a foot or two away.

101 A beam engine that operated chain haulage at 'Boggart Hole' near Townley Hall, Burnley, photographed in early 1946. The end of the beams were spherical and probably all in one piece with the pins. The connecting rod of cast iron and X-section, together with the built-up flywheel, all point to it being a very old engine. All efforts to find a photograph of it in use, or even with the house intact, have been unsuccessful.

9 A Compound Pumping Engine

Not far from Burnley, at Reedley Colliery, there was a pumping engine of some historic interest, but before describing it a few words about the type of engine will not be out of place. From early times it was known that to make a more economical engine it would be necessary to use steam at a higher pressure so that the same amount of steam would do more work. As materials and boiler-making improved it became possible to generate steam at higher pressures and engines had to be designed that could use such steam. Most Cornish engines were designed to work at about $40lb/in^2$, but in practice many worked a little higher than this. The Rockingham vertical winding engine worked at $50lb/in^2$, while the Wombwell and Diamond Pit winders used steam at $80 lb/in^2$.

Compound, and later triple-expansion, engines became popular and much theorizing has been done about the advantages of two cylinders for high and low pressure kept at different temperatures. Compound engines were probably more economical, because high-pressure steam could not leak through as easily as in an engine with one set of valves and a single piston. When it came to building compound non-rotative pumping engines there were additional

102 Hathorn Davey pumping engine at Reedley Colliery near Burnley in 1946. The inverted T-bob had a balance weight at the left, while the right-hand side drove two pump rods. The rod at the top right-hand drove another identical bob. The two objects in the foreground were probably bottom clacks rather than bucket pumps, because they do not appear to have any provision for fixing to the pump rods. Note the two extraordinarily-large split pins at the top of the bob.

103 Another view of the outdoor part of the Hathorn Davey engine. Lower down can be seen one of the pump rods hung from a stub nose-pin in the bob. An unusual feature was the hand rail fixed to the horizontal connecting rods. At the left-hand side is a cage for the pump men to ride in.

problems, such as the extra shock on pump rods that arose from higher pressure steam. One of the few, or perhaps the only, successful engine in this category was the differential compound engine invented by Henry Davey in 1871, and built by Hathorn Davey of Leeds, which used steam at up to 110 lb/in².

In this engine, which was horizontal, there was no trip gear to allow the steam valve to open suddenly and cause a shock on the pump rods, as it did in Cornish engines. In its place there was a small engine, very much like some forms of boiler feed-pump, although the pump part was a dashpot which regulated the speed of the small engine. There was also another dashpot that governed the length of the pause at each end of the stroke. This small engine could be set to work at the same rate as the main engine. If through reduced load, due to breakage or leakage, the engine tended to move faster than the small engine it automatically used less steam, or in the case of a serious loss of load, steam would be admitted to the wrong side of the piston and prevent damage. The small engine was often referred to as the 'steam man' or the 'iron man'. Like a good engine man it stood at the side of the engine. In Henry Davey's original design it was not in that position, but laid in a horizontal position parallel with the main engine valve-spindle. The model in the Science Museum, South Kensington, shows the same arrangement. When the builders changed the design is not known, but the only engine seen with the older arrangement was the one at Reedley.

With the exception of the Newcomen engine shown in Fig 27 all the other pumping engines shown had plunger pumps, but for some unexplained reason nineteenth-century colliery engineers had a preference for bucket pumps. This engine at Reedley was a strange mixture of new and old, for when it was built it was quite an advance in pumping engine design and yet the pumps were no different in principle to those used by Newcomen. Davey engines, being double-acting, were arranged with two angle bobs at the top of the mine shaft and two pump-rods. The engine at Reedley had four rods working in the shaft, each working an 18in bucket pump lifting approximately 150ft, giving a total lift of 600ft.

In the colliery office there was a miner's surveying dial of an early type and engraved around the outer edge was the owner's name: 'The exors of . . . Hargreaves'. In the days of the old colliery company this was shown to visitors. Both the pumping engine and the miner's dial looked to be older than the colliery, which produced its first coal on 13 November 1891.

At least one Cornish pumping engine was altered to Davey's differential principle. This was the Harvey-built *Jumbo* at Millclose Mine, Derbyshire, Britain's richest lead mine. This 80in engine was built in 1875 and when the differential engine was fitted it did away with the horns, handles and cataract gear, but made the engine more difficult to start. This Millclose engine was also unusual in that the pump rods were not connected directly to the bob, but to a crosshead working in guides below the top of the shaft. The crosshead was connected to the bob with two iron rods.

104 This end-on view of the Hathorn Davey engine shows how the low-pressure cylinder was placed behind the high-pressure one. To avoid the use of packing between the cylinders and the use of two covers there was a common cylinder cover and two piston rods from the rear cylinder.

105 Copy Pit at Cliviger with an unusal back-to-back winding arrangement, as it was about 1946. *(Burnley Public Library)*.

There were other items of historic interest in the Burnley Coalfield. Copy Pit at Cliviger had an unusual winding engine. The single horizontal cylinder had four valves, arranged so that the steam valves were at one side and the exhaust at the other side. A rocking shaft above the cylinder operated the valves. This was not a common way of building engines, although there were no doubt others built in a similar fashion. The rocking shaft was worked by a long handle, so that the engine could not work except when the engine man worked the handle. Other engines with hand gear have been heard of and it seems that they were used to obviate any chance of running away out of control. Perhaps the engine at Cliviger had run away at some time, because there was a link motion that required oiling, but did nothing.

Even more unusual was the strange layout of the pit shafts. One was in front of the house and the other behind. The two shafts were probably only used for men and props, as the coal came up an incline with chain haulage. One of the shafts had a pair of bucket pumps that were worked from two angle bobs at the shaft top. These bobs were not like those built by Cornishmen, for the two sides at right-angles formed a single casting, while the third side was a tie bolt or link. There did not appear to be any means of ensuring that the third side was in tension. Cornish-built angle bobs with wooden arms no doubt gave a little and ensured that the third side was in tension, but these could not give. Cast-iron bobs must have proved safe enough for others could be seen not too long ago to be remembered and yet more can be seen on old photographs. A further example of confidence in cast iron was shown in the way the angle bobs were driven by a cast double-throw crankshaft, with no bearing between the two throws. This pumping gear was driven by a single-cylinder horizontal engine through a set of gears, no doubt also cast.

No condenser was seen nor was any exhaust steam. Some colliery engines in the district exhausted up the chimneys, but in this case a flue ran up the hillside to a coffee-pot chimney, just like many old lead smelting works. Another visit, in 1948, to find if the engine had a waterfall condenser with a submerged outlet in the adit, revealed the steam plant all broken up. This was not surprising as it was completely worn out.

106 Hapton Valley Pit in 1908 after a fire. Two large angle bobs in front of the engine house show that there were pumps in the shaft. It also looks as if the ropes were wide flat ones. A spare cage at the left of the headgear is very like the one taken at Reedley Colliery nearly forty years later. *(Burnley Public Library).*

10 Ventilating Fans and Engines

Copy Pit had an underground ventilating furnace burning in 1946 and for some time afterwards. It looked strange to see a small chimney on a hillside smoking for no obvious reason. At Clifton Colliery, Burnley, there was also a ventilating furnace and an underground boiler for a haulage engine. The smoke from these came out of a chimney that was adjacent to the colliery buildings, so did not look unusual. Nobody would suspect that it was in effect a chimney over 900ft high. These outlets were known as cupolas, not only in this part, but also in South Yorkshire, where their use was discontinued many years ago, probably beyond living memory.

The use of an underground furnace for ventilating a colliery is an old idea and formerly widely used. That their use survived at Cliveger and Clifton can only be explained by the pits being non-gassy. Not only was a ventilating furnace somewhat inefficient, but positively dangerous in gassy pits. It was the danger, rather than the inefficiency, that led to efforts being made to produce mechanical ventilating means.

Some of the early ideas were displacement machines, with chambers that were driven up and down through a water seal arranged like a gas holder. Other machines had shutters working

107 Clifton Colliery. This was almost in the town of Burnly and was not very noticable because there was not much waste dumped. The headgear with small wheels had formerly been for a pumping engine. In the 1930s, when interest in early engineering was awakening, the sight of such a headgear stimulated enquiries as to whether a pumping engine existed. The low, wide stack was in fact the top of the 915ft chimney for a ventilating furnace and underground boiler for a haulage engine. This photograph was taken in 1948, less than 5 years before the colliery closed.

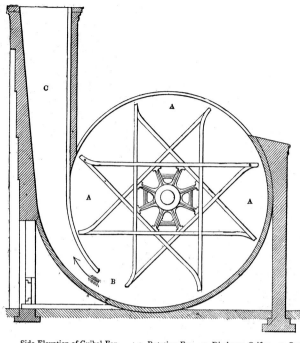

Side Elevation of Guibal Fan.   A A, Rotating Fan.   B, Discharge Orifice.   C, Outlet.

108 A Guibal ventilating fan.

in closed spaces. None of these machines could pass the amount of air required and the first successful machine was the type of fan invented in 1862 by T. Guibal. This had a large runner with a number of flat blades or wafters. These were not radial but leaned backwards. The runners were built of wood and angle iron with a central cast-iron spider. The runner revolved inside a masonry chamber with a scientifically designed shape. The inlet was at one side and rather larger than the spider, while the outlet was at one edge. An essential feature was a sliding shutter at the outlet from the fan housing into an expanding chimney or 'evasee'. The shutter was only moved if there was a big change in the load on the fan.

Most of these Guibal fans were driven by horizontal engines of a strange type. In the middle of a long bedplate there was a bearing for the crankshaft and a cylinder at each end, together with all the other essential parts of an engine. Only one end could be used at once, so it was obvious that the designer's idea was to use one half and when repairs needed doing use the other side. The weak point was that the main bearings which carried the weight of the fan were common to both halves. In any case much dirt collected in the fans so they had to be shut down at times, but even then if the facts could be known it is very likely that one of these old fans would hold the record for a long non-stop run.

The engines were made by Black, Hawthorn & Co of Gateshead, but whether the double engine was their idea or stipulated by T. Guibal is not known. It is extremely unlikely that one is still in existence now, but a few were working just after World War II when the model was built. These simple fans showed the way to improved types, such as the Walker fan which had curved blades and a notched shutter to lessen vibration. Another type was the Waddle fan that ran in the open air and discharged all round the periphery.

109 A model of a Guibal fan and steam engine. Note that this is not the Black Hawthorn 'double' engine described in the text.

110 Another view of the Guibal fan model, showing the air inlet on the right, the outlet at the top and the control for the shutter.

11 Der Fahrkunst — A Strange German Mining Machine

In previous chapters it has been explained how both water and steam power were applied to pumping, winding and crushing. Today, one can see electric power performing these functions, although it should be remembered that electricity is only a means of transmitting power that has been generated by water or some form of engine, often steam.

In a previous age both water and steam power was used to drive a strange form of mining machine that has no modern counterpart. When considering early winding gear it was mentioned that winding was often done with free hanging egg-shaped buckets known as kibbles. The name for these buckets is so much like the German *Kübel* (a tub), that one suspects that the name came from Germany. On the other hand collieries often wound coal in baskets, but all these early methods had one thing in common, they were too dangerous to wind men. Ropes could rot and chains corrode, especially in ore mines where there was acidic water. Another factor was that many metal-mine shafts were not only out of plumb, but also crooked, having been sunk on the lode. This would cause enough friction to destroy the best ropes or chains in a short time. Men used to enter and leave mines by climbing ladders. Ladder climbing was not only a waste of time, but resulted in only young men being able to work and conversely excluded older, experienced miners.

Early in the last century the deep mines in the Harz area of Germany were having considerable difficulty getting the miners to their workplaces. In 1833 Bergmeister Dorell tried out an idea at a shaft in Spiegaltal. He fitted steps and breast-high handles to the pump rods. These steps and handles were stroke-lengths apart, so that a man could stand on a step and ride up one stroke length. Just at the moment when the rod was about to reverse its movement he crossed onto a step on the other rod, then at the bottom of its movement. By repeated side-stepping, first one way and then the other way, he could reach the surface through the power of the waterwheel-driven pump rods.

Three facts emerge from this. At that time Germans must have been using plunger pumps, and not bucket pumps with the rods down the columns. Secondly they must have had the pump rods side-by-side and not with a column, or rising main, between them. Thirdly the rods could not have been entirely surrounded by guide pieces, as was the usual practice in Cornwall.

Dorell's experiment was such a success that in a few years all the deep shafts in the Harz had been fitted with a *Fahrkunst*, as these climbing aids were called. Their use spread to Belgium and France where they were called *Machine à monter.* As two reciprocating rods could be balanced by bobs or against each other, not only at the surface but at other points, it followed that no part would be in excessive tension. This led to the view that as mines got deeper some similar machine would have to be used for bringing minerals or coal up from great depths. Such a machine was actually built at Anzin in France, but little seems to have been recorded as to how it operated.

Before long the use of the Fahrkunst had spread to England where it became known as the man engine. Their use in England was limited to metal mines, and never in collieries as in Belgium and France. One slight advantage of a man engine was that men could get on or off, and travel up or down, just as their work required them to, without having to signal to an engine

111 A 'double-acting' man engine at the Grube Samson silver mine in the Harz, Germany. It was originally built in 1837 and until 1910 its length was 800m (2,600ft), when it was reduced to 200m (650ft). The steps are 3.2m (10½ft) apart and the stroke is half this. (*A. Klähn*).

112 Another view of the Grube Samson man engine. Like the flat rods seen in Cornwall there is about half a second pause at the end of the stroke, when in theory it should reverse immediately. Although the man engine was invented because ropes were not safe enough, this example has had the rods replaced by wire ropes at some time. (*A. Klähn*).

113 The all-wooden water-wheel, 39ft diameter, that drove the Grube Samson engine until 1922, since when it has been powered by electric motor.

driver at surface. Whether it was this convenience or not is not known, but man engines were also used in the copper mines in the Lake Superior area of America.

The best-known man engine is the one involved in the Levant Mine disaster in 1919 when the cap on the engine rod broke and precipitated thirty-one men to their deaths. But this was not the last man engine to work in Britain, as is widely believed. One at Laxey, driven by a water-pressure engine, worked until 1929 and even in 1974 an old resident in Laxey could tell about using it. He explained that the underside of each fixed platform had a bellmouth to prevent men being caught by their shoulders.

At Kongsberg in Norway visitors to a silver mine are shown disused machinery in an underground chamber. One of the machines is the driving gear for what they call *Fahrkunsten*. This was installed in 1879 and a short length at the top has been left intact so that it can be demonstrated. Grube Samson, at St Andreasberg in the Harz, is not many miles from where Dorell tried his experiment and today visitors can see all the surface works of an old mine. There is a reversing waterwheel for winding and a larger wheel that drove the Fahrkunst. The mine (*Grube*) closed in 1910 but all is preserved. When the mine was in use it took a man two to three hours to climb 800 metres, but with the *Fahrkunst* it took about 45 minutes. Parts of the mine have been incorporated into an electric power scheme with water turbines at 190 metres from the surface, and the *Fahrkunst* is still used by men going to attend to the underground turbines etc.

Most Cornish (and the Laxey) man engines had a single rod and fixed landings. This would be slower, but probably safer, than the German arrangement with the two rods. More likely it was a reflection on the type of pump rods used in the two countries.

What few photographs exist of Cornish man engines have been reproduced time and time again, but two models do exist which give a fairly good impression of men riding up a shaft on steps fitted to a rod. A model in Holman Brothers' museum at Camborne shows a perpendicular shaft and a typical Cornish rotary beam-engine. The other model is in Earby Mines Museum between Skipton and Colne, and shows an inclined shaft and a waterwheel instead of an engine. This model is like the other in that it has only a single rod, yet has a somewhat German air about the surface buildings.

Index